research in conservation

The Getty Conservation Institute
Los Angeles

Stone Conservation

An Overview of Current Research

Second Edition

Eric Doehne and Clifford A. Price

2010

The Getty Conservation Institute works internationally to advance conservation practice in the visual arts—broadly interpreted to include objects, collections, architecture, and sites. The Institute serves the conservation community through scientific research, education and training, model field projects, and the dissemination of the results of both its own work and the work of others in the field. In all its endeavors, the GCI focuses on the creation and delivery of knowledge that will benefit the professionals and organizations responsible for the conservation of the world's cultural heritage.

Research in Conservation
The Research in Conservation reference series presents the finding of research conducted by the Getty Conservation Institute and its individual and institutional research partners, as well as state-of-the-art reviews of conservation literature. Each volume covers a topic of current interest to conservator and conservation scientist. Other volumes in the Research in Conservation series include *Alkoxysilanes and the Consolidation of Stone* (Wheeler 2005), *Analysis of Modern Paints* (Learner 2004), *Effects of Light on Materials in Collections* (Schaeffer 2001), *Inert Gases in the Control of Museum Insect Pests* (Selwitz and Maekawa 1998), *Oxygen-Free Museum Cases* (Maekawa 1998), *Accelerated Aging: Photochemical and Thermal Aspects* (Feller 1994), and *Statistical Analysis in Art Conservation Research* (Reedy and Reedy 1988).

Published by the Getty Conservation Institute
Getty Publications
1200 Getty Center Drive, Suite 500
Los Angeles, California 90049-1682
www.gettypublications.org

Gregory M. Britton, *Publisher*

Ann Lucke, *Managing Editor*
Beatrice Hohenegger, *Project Editor*
Cynthia Newman Bohn, *Manuscript Editor*
Pamela Heath, *Production Coordinator*
Hespenheide Design, *Designer*

Printed in Canada
Library of Congress Cataloging-in-Publication Data

Doehne, Eric Ferguson.
Stone conservation : an overview of current research / Eric Doehne and Clifford A. Price. —2nd ed.
p. cm.—(Research in conservation)
Includes bibliographical references and index.
ISBN 978-1-60606-046-9 (pbk.)
1. Building stones—Deterioration. 2. Stone buildings—Conservation and restoration. I. Price, C. A. II. Getty Conservation Institute. III. Title.
TA426.P75 2010
691'.2—dc22

2010019971

Contents

Foreword to the Second Edition, 2010

Petra, Angkor, Copán, Venice, Lascaux, Easter Island. Stone conservation research may not be the first thing that comes to mind when reading these words, but it is because these places of irreplaceable cultural heritage, and many other stone monuments, are eroding at a noticeable rate that the subject of this volume is of such crucial importance. In the summer of 1994, the Getty Conservation Institute (GCI) invited Professor Clifford A. Price to provide an overview of research on the conservation of stone monuments, sculpture, and archaeological sites. The purpose of the review, which was subsequently published in the 1996 book *Stone Conservation: An Overview of Current Research*, was to inform GCI research policy in this field and to highlight areas into which Getty resources might usefully be channeled. Today, a Google search for "stone conservation" raises this book in the first link—a testament to its enduring usefulness to the wider conservation community.

Stone Conservation remains one of the most cited and downloaded of the GCI's books some fifteen years after it was written. A refreshingly opinionated work, its call to reform the focus and process of research was subsequently echoed and reinforced by other authors and institutions. By raising challenging issues, the book influenced a generation of conservators and scientists who have worked to advance the field of stone conservation. Indeed, progress on several key issues can be tied directly back to Professor Price's frank observations and prescriptions. The need for both a conservation journal that reviewed scholarly articles and for rigorous peer review of conservation publications contributed to the subsequent formation of venues such as the *IIC Journal, Reviews in Conservation*, and the expansion of the GCI's *Research in Conservation* series. His call to improve the quality and timing of conferences and the accessibility of proceedings was one of the stimuli for the recent development of the Torun Guidelines adopted at the ICOMOS International Stone Committee meeting in 2008.

The fact that the stone conservation field has evolved significantly since 1994 prompted requests to update this popular volume, and in May

2007 GCI scientist Eric Doehne, with the advice and collaboration of Clifford Price, embarked on a new survey of the field of stone conservation research. The goal was to retain key characteristics of the first edition (notably its brevity, informal character, and pointed suggestions), while covering the recent explosion of new research, enlarging the discussion of preventive conservation, and adding new sections on rock art and other subjects. This required a parallel compilation of a new bibliography, which included a review of more than six thousand abstracts and more than three thousand PDF files of material published between 1995 and 2009. Topics ranged from nano-scale measurements of salt damage by materials scientists to conservators' documenting the unintended consequences of waterproofing agents. The selected bibliography drawn from this research effort is included in this new edition as an appendix and will be a useful starting point for many researchers.

With increasing reliance on the Internet and the rapid development of interdisciplinary research and teaching, we live in a time when all knowledge is being connected to all other knowledge. Building and maintaining a coherent infrastructure for the conservation field, arguably one of the most interdisciplinary of endeavors, is a particular challenge. To advance the field of stone conservation and manage the growing variety and volume of information, practicing conservators and scientists need a framework for building a coherent base of useful knowledge. The second edition of *Stone Conservation: An Overview of Current Research* provides this framework in the form of a strategic overview of the past fifteen years in stone conservation research and an updated critique of the field's strengths and weaknesses, with recommendations for future research.

Timothy P. Whalen, Director
The Getty Conservation Institute

Preface to the Second Edition, 2010

Being a conservation scientist often means acting as a bridge—between researchers and conservation practitioners, and between the many different fields of research related to the preservation and conservation of carved and worked stone, from Stonehenge to the Sphinx, and from analytical chemistry to X-ray tomography. Like the first edition, this volume is not a literature review. It is an overview that maps the landscape of stone conservation, cites interesting and representative research, and is intended to serve as a useful point of entry to the field.

I began the research for the second edition in May 2007 as an effort to update and highlight the significant changes that had taken place in the field since the first edition and to encompass a much larger range of publications. The text was largely written in 2008, with revisions and editing completed in 2009. Such an endeavor unavoidably results in a particular perspective. This tendency has been ameliorated by consulting with an experienced, as well as linguistically diverse, group of conservation practitioners, researchers, and colleagues who have been very generous with their time.

In particular, I would like to acknowledge the help and advice given by John Ashurst, Api Charola, Jose Delgado Rodrigues, Vasco Fassina, John Fidler, William Ginell, John Griswold, Chris Hall, Seamus Hanna, Adrian Heritage, Ioanna Kakouli, Lorenzo Lazzarini, Susan Macdonald, Bill Martin, David Odgers, Leo Pel, Sarah Pinchin, Francesca

Pique, Jerry Podany, Thomas Roby, Carlos Rodríguez-Navarro, Eduardo Sanchez, Alison Sawdy, George Scherer, Stefan Simon, Michael Steiger, Marisa Laurenzi Tabasso, Giorgio Torraca, Véronique Vergès-Belmin, Heather Viles, Norman Weiss, George Wheeler, Chris Wood, Konrad Zehnder, and Fulvio Zezza. I would also like to acknowledge former GCI interns and postdoctoral researchers Enrica Balboni, Ann Bourgés, Tiziana Lombardo, Paula Lopez Arce, and Claire Moreau, who helped teach me more about stone through our joint research. The students of the International Course on Stone Conservation have also been a source of inspiration. The GCI's Beril Bicer-Simsir, David Carson, Giacomo Chiari, Mara Schiro, and Jeanne Marie Teutonico provided important support. I am grateful to Valerie Greathouse and Tina Segler for their help in tracking down references and to Cynthia Godlewski and Cynthia Newman Bohn for their excellent coordination and editorial assistance. Two anonymous peer reviews of earlier drafts of the book were thorough and thoughtful. Finally, my coauthor has been brilliant in skillfully aiding my efforts to bind together the old and the new in this volume, and I extend my kind thanks to him for his enthusiasm for this project.

Eric Doehne
Pasadena, California, June 2009

Foreword to the First Edition, 1996

The "sympathetic conservation" of historic or culturally significant stone is a relatively recently recognized practice. In the past, the repair of damaged sculptured stone objects was frequently accomplished using more intrusive means, such as iron dowels, staples, or clamps that often marred the appearance of the object and could lead to further damage. For the patching and filling of defects, lime mortar, cement, plaster of Paris, sodium silicate, and various gums and resins were used—materials no longer considered acceptable. Stone-cleaning processes involved harsh acidic treatments followed, at times, by neutralization, which resulted in the production of soluble salts that penetrated the stone and increased the potential for future salt-crystallization damage. Damaged architectural stone was either replaced or repaired with little regard to the materials' compatibility with the stone, appearance matching, or the durability of the treatment.

The unsuitability of many of these treatments encouraged research efforts to develop new materials and procedures for the preservation of stone. Over the past twenty years or so, these studies have resulted in the publication of a vast number of reports and papers, most of which were concerned with case studies and how specific stone substrates were treated. Few were accompanied by details of the research that supported the selection of the treatment method or materials used. Fewer yet were those concerned with stone-damage mechanisms or with scientific research on stone conservation processes, materials behavior, and environmental effects.

Although research has proliferated, there has not been a recent, concerted effort to evaluate the direction in which research has been progressing and whether or not the current direction is proving fruitful. Should the emphasis on stone conservation research be placed on development of new materials and new application procedures? Has there been significant work on the evaluation of the post-treatment stone property improvements? Are the methods for evaluating stone properties universally accepted? Do we need to conduct research on methods for carrying out and assessing the long-term durability of treatments? Are there problems in the process of conducting stone conservation research that bear on our ability to do the research effectively? Can these problems be defined; and, if so, what can be done to further the effectiveness of stone research? These are some of the many questions that Clifford A. Price has considered in this review of the current status of stone conservation research.

We asked Dr. Price to give us his subjective viewpoint on what is being done right, what areas of current research should be continued or accelerated, and what new directions should be addressed that would promote an increase in the effectiveness of stone conservation. In the course of preparing this review, Dr. Price has had extensive discussions with a number of active participants in the stone conservation community and what has emerged is an engaging account on whither we seem to be going and in which ways, if any, our paths should be altered.

William Ginell
Emeritus Head, Architecture and Monuments
The Getty Conservation Institute, 1996

Preface to the First Edition, 1996

This volume was written over a short period during the summer of 1994, following a systematic study of the major publications of the last five years. Inevitably, the volume reflects my own experience, expertise, and linguistic abilities. An international working party could, no doubt, have produced a more objective and comprehensive report—albeit over a longer span of time. In order that my own prejudices might not shine through too strongly, I have consulted with other conservation scientists and stone conservators, and I am very grateful for the help and advice that they have given me.

In particular, I would like to acknowledge the help given by John Ashurst, Norbert Baer, Guido Biscontin, Sue Bradley, Api Charola, Vasco Fassina, John Fidler, William Ginell, Lorenzo Lazzarini, Bill Martin, Antonia Moropoulou, Marisa Laurenzi Tabasso, Jeanne Marie Teutonico, Giorgio Torraca, and George Wheeler. I am also grateful to Sasha Barnes for the help that she has given in rooting out references and to Julie Paranics for help in the final production of the volume.

The emphasis of this publication is on stone as a material. There is little reference to mortars, and no consideration of the structural performance of stone masonry. This volume is not a detailed, state-of-the-art review, and many of the references I have given are intended as illustrative rather than definitive. It is intended to give a strategic overview of the whole field and to identify areas of strength and weakness where further research should be focused.

Clifford Price
London, 1996

Dedication

We dedicate this book to the memory of John Ashurst, 1937–2008, in recognition of his unparalleled contribution to stone conservation through research, practice, and training.

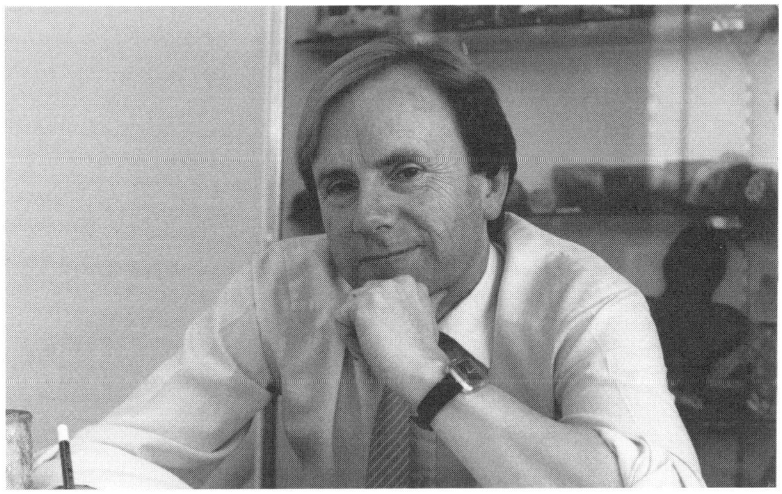

Introduction

After presenting his work at a recent stone conservation conference, a thoughtful researcher was responding to questions from conservators. A pained expression was evident on his face as he said to the audience, "I feel as though I am explaining in great detail why I cannot help you."

This encapsulates the frustration felt by many who are involved in stone conservation at present. While great strides have been made in understanding why stone decays, the perception is that much less progress has been made in helping conservators cope with a number of long-standing conservation problems. The researcher's comment also highlights a central paradox in conservation: while progress is necessarily incremental, time and the elements steadily take their toll on cultural heritage, and the window for action to ensure that history is preserved for future generations is limited.

This volume takes a broad and sometimes critical look at the present state of stone conservation and of the way in which research is conducted. It looks first at the deterioration of stone and ways in which deterioration may be prevented or remedied. Then, it discusses some of the factors that limit the effectiveness of research and makes recommendations as to how research might be made more effective. It concludes with some reflections on changes that have taken place over the past fifteen years.

Chapter 1
Stone Decay

The deterioration of stone is all too familiar to anyone who has looked closely at a historic stone building or monument. While there are a few stones that seem little affected by centuries of exposure to the weather, the majority of stones are undergoing gradual and episodic deterioration. This may not matter much if the stone is an undecorated part of a massive wall. However, it does not take much deterioration of a carved piece of stone for the sculptor's original intention to be lost altogether. A high proportion of the world's cultural heritage is built of stone, and it is slowly but inexorably disappearing.

If we are to do anything to reduce or prevent this loss of our heritage, we must first be able to characterize the many stones involved. We need to be able to describe the decay and to measure its extent, severity, and rate. We then need to understand the causes and mechanisms of decay. Only then can we hope to understand the behavior of any particular stone in a given environment.

CHARACTERIZING THE STONE

The literature is full of papers concerned with stone characterization. Pick up any set of conference proceedings and you will find numerous papers that first describe the situation and history of some particular monument and then lay out the physical properties of the stones involved. There will be petrological descriptions, followed by measurements of surface hardness, porosity, capillarity, hygric and thermal expansion, pore size distribution, mechanical strength, velocity of sound, resistance to salt crystallization, and so forth. There will invariably be photographs taken on a scanning electron microscope and, for good measure, probably some energy-dispersive X-ray analyses. To what end? The information will no doubt be of value to those who are concerned with the care of that particular monument, but it is of questionable relevance to a wider audience unless the properties of the stone can be linked to its performance. At this point the field needs to move beyond basic characterization to a better understanding of material behavior (Torraca 2009) and the maintenance necessary to sustain long-term performance (Brand 1995).

Most of the techniques for characterization are well established. Many are summarized by Robertson (1982) as well as Borelli and Urland

(1999) and Svahn (2006). Adams and MacKenzie (1998) provide a useful atlas of petrographic sections, while a more recent petrographic atlas and applications of polarized light microscopy to building materials conservation are presented by Bläuer and Kueng (2007) and Reedy (2008).

In the process of characterizing stone, it is important to recognize that while some stones have a similar composition, their behaviors may have few things in common. For example, Istrian stone, Lecce limestone, and Carrara marble are all carbonate materials, but their contrasting modes of deterioration depend more on their porosity, pore shapes, pore size distribution, and grain size than their chemical composition. One division of stone types is based on the percentage and relative ratio of pore-shaped and fissure-shaped voids (Croci and Delgado Rodrigues 2002). A second division can be made on the basis of the degree of hygric swelling of the stone (Delgado Rodrigues 2001; Duffus, Wangler, and Scherer 2008), and a third division on the strength (Winkler 1985; Bourgès 2006). Subsequent divisions based on composition, texture, and homogeneity enable further distinctions to be made, but they may be less important in rating overall performance than the first three. Those stones with high porosity, high rates of swelling, and low strength tend to be relatively poor building materials (e.g., Jackson et al. 2005).

A review of the relationship between pore structure and other stone characteristics is given by Bourgès et al. (2008). Gauri and Bandyopadhyay (1999) review the interpretation of mercury porosimetry data and cite a number of the seminal papers on pore structure determination. Analysis of the positive correlation between the fractal dimension, stone pore surface, and the degree of natural weathering has shown that increases in the surface fractal dimension are a more accurate descriptor of the degree of weathering than pore size distribution (Yerrapragada, Tambe, and Gauri 1993; Pérez Bernal and Bello López 2000).

DESCRIBING DECAY

Stone decay takes many different forms. Stone may weather away gradually, leaving a sound surface behind; at times large scales of stone may drop away in one episode. Sometimes the surface erupts into blisters; sometimes the stone loses all integrity and simply crumbles away. Sometimes the stone may look perfectly sound to the naked eye, while below the surface it has lost its cohesion.

One of the problems inherent in discussing stone decay is finding a common language. Even in English, there are a bewildering number of terms that may mean different things to different people. And even if we can agree on terms to describe the types of decay that we observe, it can be difficult to determine the severity or rate of decay. A significant advance in this area is the recent publication of a stone decay glossary by the ICOMOS Stone Committee under the editorship of Véronique Vergès-Belmin (2008). Another effort to produce a glossary of decay terms is that of the Italian Commissione NORMAL (UNI 2006). Earlier work in this area came from the building stone industry in an effort to standardize terminology (Stone Federation of Great Britain 1991), governmental organizations (Grimmer 1984), and research groups (Fitzner, Heinrichs, and Kownatzki 1997).

The ICOMOS-ISCS *Illustrated Glossary on Stone Deterioration Patterns* (Vergès-Belmin 2008) helps define and clarify usage across languages and within the stone community, providing useful definitions of terms such as scaling, spalling, and flaking. Weathering is generally defined as the result of natural atmospheric phenomena, while decay is "any chemical or physical modification of the intrinsic stone properties leading to a loss of value or to the impairment of use," degradation is "decline in condition, quality, or functional capacity," and deterioration is the "process of making or becoming worse or lower in quality, value, character, etc." Some interesting details of the history of stone glossaries can be found in the introduction to the glossary.

A more guided approach than a glossary can be found in work on expert systems from the late 1990s, with Van Balen (1996; 1999) producing an atlas of damage to historic brick structures as part of an expert system for elucidating environmental effects on brick. The atlas evolved into a broader program known as the MDDS (Masonry Damage Diagnostic System) (Van Hees, Naldini, and Sanders 2006; Van Hees, Naldini, and Lubelli 2009). Expert systems have gone in and out of fashion over the past fifteen years, but the need to capture expert experience and judgment has become ever more urgent, given the large number of conservation professionals nearing retirement age.

Fitzner has produced an important classification of weathering forms as a basis for mapping the deterioration across a building facade (Fitzner, Heinrichs, and Kownatzki 1997). This system has also been presented in case studies (Fitzner, Heinrichs, and La Bouchardiere 2004). Such complex systems have been criticized because of the number of parameters to be measured (Moraes Rodigues and Emery 2008) as well as "cost concerns and the extensive training they require" (Dorn et al. 2008). Fitzner's classification recognizes nineteen different weathering forms and goes some way toward recording the severity of each, based on visual inspection (Fitzner 2004). Similar, but simpler systems have been described by Massa, Naldini, and Rorro (1991) and by Vergès-Belmin (1992). Zezza (1990; 1994; 2002) has used digital image processing to map different forms of surface decay. Starting with photographs and other nondestructive information, such as ultrasonic measurements, false color images are produced that identify particular forms of decay.

HOW SERIOUS IS IT?
MEASURING THE EXTENT AND SEVERITY OF DECAY

In order to make real progress, we need to quantify decay. In other words, in addition to describing the type of decay, it is essential that we are able to measure its extent, or the area it covers; its severity, or how advanced the decay is; and the rate of decay over time. First, we need to do so in order to unravel its various causes. For example, how can we say that pollution is causing decay unless we have some way of correlating pollution levels with decay? Second, we need to have some objective means of assessing the extent and the rate of decay in order to decide whether remedial action is necessary and, if so, how urgent the need is. Third, we cannot establish whether our remedial actions are having any effect unless we can monitor the condition of the stone afterward.

If one accepts these eminently reasonable preconditions, then we are left with a situation where extremely few monuments today (or even paintings) meet these basic conditions. Conservation decisions most often rest upon a framework of experience and general guidelines for treatment compatibility, instead of data on the actual behavior or rate of loss of the monument. Conservation documentation for the majority of our cultural heritage appears to consist of a few uncalibrated photographs taken under different lighting conditions over a few decades. Helping to fill this void with more quantitative and reproducible approaches has been the objective of many of the research projects cited in this volume: turning "weathering" or "decay" into numbers.

No single technique is sufficient to measure stone deterioration, since decay takes many different forms. Some techniques, such as 3D laser scanning and fluorescence LIDAR (light detection and ranging), look only at the surface, and they are well suited to decay that consists of a gradual loss of surface, leaving sound stone behind. Other techniques, such as ultrasonic measurements, thermography, or magnetic resonance imaging (MRI) are designed to probe below the surface, and these are useful where decay consists of a loss of cohesion within the stone, or the development of detached layers, blisters, or internal voids.

Before using more complex methods, simple visual examination plays an important role in quantifying decay. A single examination can convey the state of the stone at a particular moment, but it does not capture the rate of decay. For this, a series of inspections is required, usually over a period of several years. Photographs are of immense value here, but their objectivity can be abused. Winkler (1975, p. 87), for example, constructs an alarming graph of exponentially increasing decay on the basis of just two photographs. Even within a series of photographs, a fundamental difficulty is that often they have been shot under differing lighting conditions, making the interpretation of surface loss challenging (GCI and IHAH 2006; Thornbush and Viles 2008).

Two improvements in traditional photographic documentation show promise. One is the use of time-lapse methods to provide more frequent images in order to correlate surface loss with environmental changes (Sawdy and Heritage 2007; Doehne and Pinchin 2008; Zehnder and Schoch 2009). The other is a new method known as polynomial transform mapping (PTM), a subset of RTI (Reflectance Transform Imaging), that is, the use of multiple photographs from different angles to document more comprehensively the texture of stone surfaces. This gives the viewer the ability to control the angle of the light source in a given image using Java-based software (Malzbender, Gelb, and Wolters 2001; Padfield, Saunders, and Malzbender 2005). See examples at: http://www

.hpl.hp.com/news/2004/jan-mar/ptm.html and http://www.southampton.ac.uk/archaeology/acrg/acrg_research_PTM_amazon.html.

Surface Techniques

Surface techniques for quantifying rates of stone loss include the use of a microerosion meter, profilometry, close-range photogrammetry, laser scanning, and laser interferometry. The microerosion meter is a simple micrometer device that measures surface height at a number of predetermined points relative to datum studs set into the stone. It was used, for example, to monitor the rate of stone decay at St. Paul's Cathedral, London, over a twenty-year period, during which atmospheric sulfur dioxide levels in the region fell by 50 percent (Trudgill et al. 1992; Trudgill et al. 2001). Erosion rates on horizontal sites were found to have decreased from 0.045 mm/year in the period 1980–90 to 0.025 mm/year in 1990–2000.

Optical profilometry is a contact-free technique that consists of the projection of a grid of lines onto the surface at an angle of 45°. Any irregularities in the surface are immediately evident. Aires-Barros, Maurício, and Figueiredo (1994) have demonstrated its use, coupled with image analysis, to construct a weatherability index. Similar optical methods include laser triangulation, confocal microscopy, and digital holography.

A technique for monitoring surface roughness known as contact profilometry was utilized by Jaynes and Cooke (1987) to monitor the decay of limestone when exposed to a range of different pollution environments. It measures irregularities by means of a stylus that is drawn across the surface; movement of the stylus produces an electrical signal in a transducer.

Grissom has compared stylus profilometry, reflected-light image analysis, and visual/tactile evaluation to assess the roughness of abrasive-cleaned stone. The results found tactile evaluation to be the "more practical and cost-effective technique" (Grissom, Charola, and Wachowiak 2000).

Close-range photogrammetry was described by Coe and others (1992), who demonstrated that the technique was sufficiently sensitive to detect surface loss of 0.1 mm per year over a four-year period. More recent work has pointed out the importance of human interpretation in close-range photogrammetry (Inkpen, Collier, and Fontana 2000) and has shown how to combine laser scans with close-range photogrammetry (Ressl 2007).

Asmus and co-workers (1973) were among the first to propose the use of laser interferometry to monitor surface loss in stone. The technique has now been developed to the point where deformations as small as 0.5 microns can be detected. Laser profilometry has also been used to quantify changes in surface roughness due to laser cleaning (Colombo et al. 2007). Meinlschmidt and others (1992; 1998) have demonstrated the use of a portable system based on electronic speckle pattern interferometry (ESPI) or video holography. They were able, for example, to monitor deformations that took place during the hardening of a mortar or the growth of efflorescence over a period of just a few days. Recent advances have made such systems less expensive and more practical in field conditions (Keene and Chiang 2009).

Many stone decay processes can be evaluated by focusing on solution chemistry and mineral reactions. Microcatchment studies are a useful way to evaluate the chemical dissolution of stone surfaces, where the ions in rain runoff are measured to evaluate reaction rates (Halsey 2000). Finally, atomic force microscopy (AFM) and vertical scanning interferometry (VSI) have been used to monitor mineral reactions and the effects of biodeterioration (Davis and Lüttge 2005; Perry, McNamara, and Mitchell 2005; Herrera, Le Borgne, and Videla 2009).

Looking Beneath the Surface

Outward appearances may be sufficient in some instances, but they can be deceptive. It is not unusual to find a stone surface that looks perfectly sound but which sounds hollow when tapped. Sooner or later, we need a way of measuring what is going on beneath the surface.

Many techniques are available and some of the more important are reviewed by Facaoaru and Lugnani (1993). These are typically divided into in situ field methods and laboratory-based methods. Lab tests are performed on collected samples or on samples subjected to accelerated or artificial weathering.

In Situ Field Methods

Preeminent among field methods is the use of ultrasonics to detect the presence of cracks, voids, and other inhomogeneities in stone (Mamillan 1991; Bläuer Böhm 2004). This may take a variety of forms, such as using the longitudinal wave or the transverse component running parallel to the surface. Galán and co-workers (1992) provide an early case study demonstrating the reliability and cost-effectiveness of the technique.

The transmission of ultrasonic waves in stone depends on many factors, and interpretation of the data is not necessarily straightforward. Valdeón, King, and De Freitas (1992) used digital analysis of the surface wave to demonstrate that wave attenuation can provide a sensitive measure of stone decay. The velocity of the longitudinal wave was a less sensitive measure. Montoto, Valdeón, and Esbert (1996) have used ultrasonic tomography to investigate the internal deterioration of megaliths in northwestern Spain. The technique was useful for determining the position of internal fissures but was less reliable at assessing the condition of stone immediately below the surface. Simon and co-workers (1994) have used formal concept analysis to optimize the interpretation of ultrasonic velocity measurements, while Mosch and Siegesmund (2007) correlated a large data set of physical stone properties with ultrasonic measurements. Weiss, Rasolofosaon, and Siegesmund (2002) found ultrasonic measurements a useful method to measure degradation of marble from thermal cycling. However, because the presence of moisture can produce misleading results, it is critical that the marble be dry before ultrasonic measurements are taken to ensure consistent results (Siegesmund, Weiss, and Rüdrich 2004). Ultrasonic testing has also been found useful for addressing the difficult challenge of long-term evaluation of stone treatments (Simon and Lind 1999; Favaro et al. 2006; Favaro et al. 2007).

The development and application of the drilling resistance measurement system (DRMS), also known as the drilling force measurement system (DFMS), has provided an extremely useful and minimally destructive method for evaluating the condition of stone and the performance of treatments for stone (Lotzmann and Sasse 1999; Leroux et al. 2000; Delgado Rodrigues, Ferreira Pinto, and Rodrigues da Costa 2002; Pamplona et al. 2008). The system uses a portable drill and ceramic drill bit with a sensor to measure the force needed to advance the drill bit a given distance. In principle, the DFMS can determine depth of deterioration and the penetration depth of consolidants, in situ, with a minimum of destruction (a 3 mm hole).

Ground-penetrating radar is increasingly used in archaeological prospecting, and it is natural that its use should be extended to historic buildings (Finzi, Massa, and Morero 1992). It has seen wider application recently by a number of researchers (Binda et al. 2003; Binda, Lualdi, and Saisi 2007; Huneau et al. 2008; Palieraki et al. 2008). The method is useful in detecting flaws, voids, moisture, metal straps, and the thickness of stone masonry.

Infrared thermography has been used by a wide range of researchers (Moropoulou, Avdelidis, and Theoulakis 2003; Grinzato et al. 2004; Tavukçuoglu et al. 2005) to study moisture in stone. To provide useful results, a thermal contrast, such as solar heating or deliberate heating by infrared lamps, is often needed to identify the different surface temperatures related to differences in moisture content. This method is known as photothermal radiometry and has been developed to detect delaminations and voids (Madrid, Coffman, and Ginell 1993; Candoré et al. 2008). Most building materials have significant thermal inertia, and practitioners using thermography on a casual basis will not necessarily find useful temperature contrasts.

One interesting way to look into a stone's subsurface for small surface detachments is to use the sensitivity of laser holography interferometry in combination with varying sound vibrations from a loudspeaker to map detached segments of wall paintings or stone surfaces (Castellini et al. 2003; Gulker, Hinsch, and El Jarad 2004; Keene and Chiang 2009).

A range of field evaluation methods has proven useful for quantifying water uptake and surface coherence, including the sponge and Scotch tape tests (Urzï and De Leo 2001; Vergès-Belmin and Laboure 2007; Vandevoorde et al. 2009).

Laboratory-Based Methods

So far, we have considered minimally destructive techniques. It is, of course, possible in some instances to remove samples for analysis in the laboratory. These will often consist of core samples, which are sliced parallel to the original surface. The slices may be examined using the normal techniques for characterizing stone, such as polarized light microscopy, scanning electron microscopy combined with energy-dispersive spectroscopy, hygric tests, and biaxial flexural strength measurements (Mamillan 1991; Snethlage, Wendler, and Sattler 1991; Bläuer Böhm 2004). Surface hardness measurements may be useful, and the salt content of the slices may also be determined (Bläuer Böhm 2005).

The European Commission (EC) projects COMPASS and DESALINATION have developed a simple test for salt content based on the hygroscopic moisture content (HMC) (Gonçalves and Delgado Rodrigues 2006; Gonçalves, Delgado Rodrigues, and Abreu 2006; Nasraoui, Nowik, and Lubelli 2009). Kaminski (2008) has proposed an alternative gravimetric system and makes some constructive criticisms of common aspects of the diagnosis and analysis of moisture and salts, including misleading readings from moisture meters based on electrical resistance or dialetric properties, dry drill powders showing lower than expected results, and salt solution–conditioned chambers not providing consistent conditions for HMC measurements.

Jacobs, Sevens, and Kunnen (1995) proposed the use of computerized X-ray tomography (CT) to gain further insight into the internal structure of stone and the changes that occur during the deterioration of building materials. Procedures were developed to bring the resolution down to grain-size level (about 100 microns or less). Mossotti and Castanier (1990) used CT scanning to show that for Salem limestone, capillary water reached the surface except under windy conditions, when the air/water interface moved into the stone. In the past decade, resolution of the CT method has advanced significantly (Bugani et al. 2008; Cnudde et al. 2009; Ruiz de Argandoña et al. 2009); however, it is still difficult to see treatments and salts inside pores owing to the lack of contrast and the small amount of material scanned. A promising way to overcome the limitations of x-ray CT is the use of high-speed neutron tomography (synchrotron radiation) for *in situ* dynamic analysis of wetting/drying, moisture transport, salt development, or the curing and evaluation of protective coatings and consolidants within a porous stone (Vlassenbroeck et al. 2007; Cnudde et al. 2008).

The linear variable differential transformer (LVDT; also known as a linear velocity displacement transducer) has also proven to be an important lab tool in the quantitative evaluation of the thermal and hygric response of building materials to wetting and humidity cycles, measuring expansion and contraction behavior on a micron scale (Martin, Röller, and Stöckhert 1999; Lombardo, Doehne, and Simon 2004; Poupeleer et al. 2006).

Nuclear magnetic resonance (NMR) imaging (also known as MRI) of building materials has advanced rapidly in the last fifteen years. This method allows the measurement of hydrogen and sodium ions in solutions present inside porous materials and has provided an important new dimension for understanding the behavior of moisture in building materials, especially at the millimeter to centimeter scale (Pel, Kopinga, and Brocken 1996; Pel, Huinink, and Kopinga 2002; Rijniers et al. 2004; Huinink et al. 2006; Gonçalves, Pel, and Delgado Rodrigues 2009). For example, when a stone has finer pores than an overlying plaster, NMR has shown that if both layers are fully saturated with water at the start of a drying experiment, the stone will dry after the plaster and soluble salts in the stone will tend to be retained.

All the Information We Need?

With such sophisticated forms of investigation being pursued, one might be forgiven for thinking that no problems remain in the measurement of stone decay. There is, however, a long way to go. Stone decay is a complex phenomenon, and no single technique can disentangle and quantify its causes and effects. Advances in experimental work, field measurements, and theory—each building on the other—are still needed. The techniques that we have looked at thus far are certainly useful, but the methodical measurement of decay and our understanding of decay processes over time have not yet met the goal set forth earlier of conservation decisions being based on measurements instead of assumptions.

CAUSES OF DECAY

Before we can take any action to prevent or to remedy the deterioration of stone, we must understand what is causing that deterioration. Sometimes the cause is obvious; sometimes there may be several different causes acting at once. In an attempt to clarify the relative importance and interdependency of individual causes, Verdel and Chambon (1994) have introduced the principles of system dynamics.[1] Stone decay mechanisms and rates are reviewed in the proceedings of two Dahlem meetings (Doehne and Drever 1994; Viles 1997), and both reports point out areas where additional research is needed, essentially providing useful road maps for research. An interesting example of quantifying the relative importance of a range of factors—in this case for absorption and desorption of moisture—is the careful research by Sawdy (1995; 2002). She found, for example, that for environmental control of salt decay in wall paintings, relative humidity (RH), airflow, substrate type, and temperature are important factors, while earlier research had emphasized only RH.

Some of the causes of stone decay are sudden and rapid in their effect. Those toward the latter part of the following list are slow and more insidious: earthquake, fire, flood, terrorism, vandalism, neglect, tourism, previous treatments, wind, rain, frost, temperature fluctuations, chemical attack, salt growth, pollution, biodeterioration, intrinsic factors, and so on.

The literature includes many papers dealing with the causes of decay and some reviews are available, e.g., Ashurst and Dimes 1998; Honeyborne 1998; Grassegger 1999; Feilden 2003; Smith, Gómez-Heras, and McCabe 2008. Goudie and Viles (2008) trace the remarkable history of the study of physical, chemical, and biological weathering. Recent literature is dominated by three topics: air pollution, salts, and biodeterioration. These are considered in the following sections.

Air Pollution

Air pollution is, to the minds of many, the prime culprit in stone decay. Everyone has heard of acid rain, and it is easy to conjure up an image of old buildings slowly dissolving in the rain. Needless to say, the true situation is a good deal more complex, as reviews of the role of air pollution and soiling in stone decay have found (Charola and Ware 2002; Mitchell and Searle 2004; Brimblecombe and Grossi 2007; Siegesmund, Snethlage, and Ruedrich 2008). The emphasis of these studies has largely been on limestone, marble, lime mortars, and carbonate-cemented sandstones, as these are the most vulnerable to acidic pollution. However, soiling from atmospheric particulates is a universal problem for all types of stone.

Until recently, all the attention was given to the direct effects of air pollutants on stone, and research focused on the "traditional" pollutants: sulfur oxides, nitrogen oxides, and carbon dioxide. All are naturally occurring, although human activity has greatly increased the amounts that are to be found in urban areas, as well as significantly increasing background levels of pollution in rural areas. All are capable of dissolving in water to create an acidic solution and so are capable of reacting with calcareous materials.

The direct effects of air pollution on stone received enormous attention from the mid-1970s to the early 1990s. This is due, at least in part, to concerns about the effects of pollution on health, agriculture, and the environment. Stone research in Western Europe and the United States was able to ride on the back of these concerns and to benefit from the funding of large research programs.[2]

Since the early 1990s, priorities have shifted as progress has been made in reducing sulfur dioxide (SO_2) levels in major metropolitan areas in Western Europe and the United States. Consequently, funds for research on air pollution on stone have steadily decreased and a number of larger programs have been discontinued altogether. Infrastructure and research groups, originally dependent on these large programs for the development of laboratories and funding for students, must now try to survive where there is no longer any state-supported program of research. Germany, for example, has had no federal support for stone conservation research since 1998.

In spite of the funding decrease, several conferences over the past decade have addressed important open questions regarding the impact of air pollution on rates of stone soiling and decay. One grew out of a EC-funded project[3] (Saiz-Jimenez 2004). Another set of conferences grew out of SWAPNET (Stone Weathering and Atmospheric Pollution Network), a group of researchers focused on the topic of stone decay in polluted environments that started meeting at University College London in the late 1980s. Since 1993 SWAPNET has held twelve meetings, mostly in the UK (Jones and Wakefield 1999; Mitchell and Searle 2004; Smith and Warke 1996). The most recent SWAPNET meeting was in Malta in 2007 (Gómez-Heras 2007), which reported progress in understanding the rapid decay of certain stones affected by air pollution.

Damage to stone by air pollution is still an important problem in parts of central Europe, China, India, Russia, and other industrialized regions (Larssen et al. 2006). For example, while scrubbers were installed to reduce SO_2 near the Taj Mahal, a lack of water, power outages, and the corresponding use of diesel generators were found to reduce the effectiveness of the scrubbers and decrease air quality near the site. This outlines the importance of infrastructure development to monument health (Hangal and Harwit 1997). In the past decade the rapid development of India and China's economies has in some measure been built on burning coal. This raises concerns for human health and corresponding concerns for well-known monuments, through both the direct and indirect effects of pollutants (Xingang Liu et al. 2008; Thakre, Aggorwal, and Khanna 1997; Zhao et al. 2007).

There is a general perception that air pollution is a modern problem, but Brimblecombe (1992; 2001) has shown that it is a problem that dates from antiquity. By examining the effects of pollution on individual historic buildings over periods of several hundred years, he has also attempted to correlate pollution levels with observed damage. This links in with another widespread perception: that decay rates are accelerating rapidly, despite falling levels of several major pollutants. There are insufficient data to prove conclusively whether this is indeed the case. It is possible that the perception is due largely to an increasing public awareness of the problem and to the fact that stone loss through pollution is cumulative. Also, the reaction products of air pollution, such as soluble salts, often remain on sheltered stone surfaces and result in ongoing damage.

The direct effects of acidic pollutants on calcareous stones depend very much on the immediate environment of the stone. If the stone is in an exposed position where it is regularly washed by rain, the reaction products are washed away and the surface of the stone gradually recedes. If, however, the stone is in a relatively sheltered position, the reaction products accumulate and may form dense black crusts on stone surfaces.

A great deal of research, particularly in Italy, has been concerned with the nature and the origins of the black crust (Camuffo et al. 1982; Del Monte 1992; Fassina 1992; Ausset et al. 1992). These studies have shown that carbonaceous particulate pollution resulting from the combustion of fossil fuels in electrical power generation is responsible for the blackness of the crust. More important, however, is the discovery that the particles are not passive prisoners in the crust: they contain metal oxides that catalyze the oxidation of sulfur dioxide and hence promote formation of the crust in the first place (McAlister, Smith, and Török 2008). While greater attention has now been paid to treating or removing these crusts, current research on black crusts has largely confirmed the earlier studies. Isotopic analysis has been useful in isolating the generally anthropogenic sources of black crust (Přikryl et al. 2004; Vallet et al. 2006; Siegesmund et al. 2007).

In an attempt to clarify the growth mechanism, Schiavon (1992) has studied the "stratigraphy" of black crusts. He concludes that growth occurs in two directions: inward and outward with respect to the original stone surface, but with inward growth predominating. Vergès-Belmin (1994) proposed a three-step process to explain the inward and outward formation of the black crust, making a distinction between a clear gypsum layer, growing inward through pseudomorphic replacement, and a dark one, which is a deposit, thus developing *on* the surface of the stone and growing outward. Work by Toniolo, Zerbi, and Bugini (2009) divides black crusts into three types, with marble substrates exhibiting differential preservation beneath each type.

While the vast majority of research on black crusts has focused on carbonate stone, interesting research on silicate stones has found high sulfation rates associated with diesel soot and accelerated rates of granite kaolinization associated with black crusts (Simão, Ruiz-Agudo, and Rodríguez-Navarro 2006; Schiavon 2007). In certain cases, black crusts forming on granites appear to be geochemically unrelated to the substrate and are thus accumulated from atmospheric (Silva et al. 2009) and biogenic sources (Aira et al. 2007).

Diakumaku and others have observed that some black fungi produce small spherical particles that might, under some circumstances, be confused with fly ash (1995). Microflora are also capable of producing sulfates. In the opinion of these authors, the formation of some black crusts in unpolluted environments may be attributable to biological factors. In addition, Ortega-Calvo and co-workers (1995) have demonstrated that sulfate crusts may provide an ideal habitat for some cyanobacteria through the gradual dissolution of the sulfate. Work by Mansch and Bock (1998) found greater concentrations of nitrifying bacteria and greater stone decay rates associated with air pollution and black crusts. Gonzales-del Valle and others (2003, 219) have found that "building stones host an active microflora that degrades fossil fuel derivatives." Schiavon, Chiavari, and Fabbri (2004) found organic compounds, such as polycyclic aromatic hydrocarbons (PAHs), which appear to represent markers for present-day vehicular pollution, on the limestone walls of Emmanuel College, Cambridge, UK. It seems likely that the accumulation of PAHs (Hermosin, Gaviño, and Saiz-Jimenez 2004) provides a food source for microbial communities (Saiz-Jimenez 1995; also see biodeterioration section). Despite the extensive research already carried out on black crusts, our understanding is not yet complete.

Another important area of research has been centered on the rate of decay attributable to pollutants and on the likely effect of reductions in pollution levels—a crucial issue for policy makers. For example, what would be the overall savings in building maintenance costs if sulfur dioxide levels were reduced by 10 or 20 percent? Several authors have addressed this issue through the use of damage functions, which are mathematical expressions that attempt to express the rate of stone decay as a function of several different variables. Although the damage functions differ in detail, a fairly consistent overall picture emerges (Lipfert 1989; Reddy 1990; Benarie 1991; Butlin et al. 1992; Livingston 1992; Webb et al. 1992; Meierding 1993; Livingston 1997; Viles et al. 1997; Schreiber and Meierding 1999). The rate of decay depends largely on three factors: pollution levels, rainfall acidity, and amount of rainfall.

Some intriguing findings have led to a better understanding of the interplay between material, environment, and weathering rates: for example, tropical weathering has been found to be less detrimental to marble tombstones than an acidic, polluted atmosphere (Meierding 1993, 2000).

Some authors argue that sulfur dioxide levels in certain cities have decreased to the point where sulfur dioxide is no longer a major contributor to decay. In other words, there may be a "safe level" of around 30 $\mu g/m^3$, below which sulfur dioxide is not a significant contributor to decay (Sharma and Gupta 1993).[4] This view is not universally upheld, with some experts finding that for many pollutants there is no safe threshold and that resulfation of cleaned monuments is proceeding apace in some places.[5]

One area where consensus is emerging is in the relative importance of wet and dry deposition. Where sulfur dioxide levels are high (urban areas), dry deposition appears to predominate on vertical surfaces. On horizontal surfaces and in rural areas, wet and dry deposition may be of comparable importance (BERG 1989; Butlin 1991; Furlan 1992; Cooke and Gibbs 1993; Charola and Ware 2002). According to Grossi and Murray (1999), stones with a high specific surface area and/or a deliquescent salt content were found to promote more nitrogen oxides (NO_x) dry deposition.

Some recent findings concerning the effects of air pollution have been unexpected, such as the observation of a decrease in dissolution from stone surfaces blackened with diesel soot as measured in microcatchment studies, apparently due to a higher mean surface temperature resulting in faster drying (Searle and Mitchell 2006). This counterintuitive result suggests the importance of "time of wetness" in the damage to stone, as confirmed by earlier research (Charola and Ware 2002). Recent decay of marble in New York was evaluated using mass balance methods sensitive enough to detect a 2 nm surface loss (Livingston 2008). Dissolution was found to be mostly due to gypsum dissolution originating from dry deposition with less contribution from karstic processes due to carbonic acid or from neutralization of acid rain. This result is in contrast to earlier US National Acid Precipitation Assessment Program (NAPAP) studies, which found carbonic acid responsible for approximately 70 percent of carbonate dissolution (Baedecker and Reddy 1993).

Despite significant cleaning campaigns in many European capitals, the soiling rates of historic structures remain high, apparently due to a substantial increase in diesel emissions (Grossi et al. 2003; Searle and Mitchell 2008). Recommendations for human health and monument health include increasing the distance between diesel emissions and important sites such as schools and monuments (Nord and Holenyi 1999; Sagai, Furuyama, and Ichinose 1996; Qinghua Sun et al. 2005).

What are the issues that have still to be addressed? They include the following:

- What is the role of high nitrogen oxide levels on stone decay? The substantial increase in vehicular emissions of nitrogen oxides (NO_x) has not resulted in acid attack on the same scale as SO_2. Yet despite several studies, the situation is still unclear. Some authors have found synergistic effects for NO_x on SO_2 reactions, while others have not (Kirkitsos and Sikiotis 1995; Sikiotis and Kirkitsos 1995; Massey 1999; Searle and Mitchell 2006; Allen 2007; Metaxa et al. 2009). It seems part of the issue is that a catalytic effect for NO_x on SO_2 is present in dry conditions, but not in wet (Bai, Thompson, and Martinez-Ramirez 2006). In a larger context, research is showing that the geochemical cycle of nitrogen is being altered in ways, including the impact on historic stone, that are still poorly understood (Gruber and Galloway 2008).

- What is the mechanism by which sulfur dioxide is oxidized to produce sulfuric acid? Does oxidation take place before the pollution reaches the stone, or is it catalyzed by other pollutants on the surface of the stone? Is the oxidation catalyzed by other air pollutants, such as ozone, nitrogen oxides, or diesel soot (see, for example, Rodríguez-Navarro and Sebastian 1996)? Do bacteria in the stone play a part?
- To what extent are today's decay rates influenced by pollution levels of the past (the memory effect)? For example, sulfate and nitrate salts that are already present in the stone will continue to cause damage even if further pollution were eliminated altogether. The "memory effect" story is not yet complete (Vleugels, Dewolfs, and Van Grieken 1993; Přikryl and Smith 2007).
- Recent research has examined the role of formates, acetates, and airborne microbes (Kumar et al. 1993; Grossi et al. 2003; Gibson et al. 2005; Maruthamuthu et al. 2008). Are other important pollutants being overlooked?
- What are the relative roles of carbonic acid versus sulfuric, nitric, or other acidic species? This is an issue that still remains controversial (Baedecker and Reddy 1993; Charola and Ware 2002).

More recently, the focus has shifted away from the direct effects of pollutants to their indirect effects. Carbon dioxide, generally regarded as a minor culprit where direct effects are concerned, now takes center stage. It is regarded as the primary cause of climate change, and the impact of climate change on the built heritage may far exceed the direct effects of pollutants—severe though they may be.

International concern over climate change and global warming continues to grow. Because the impact on people is the primary concern, it is easy to think that stone monuments will be immune to global warming of just a few degrees. This is not the case, however, and recent studies have started to demonstrate the widespread impacts of climate fluctuations such as floods, droughts, and humidity cycles (http://noahsark.isac.cnr.it/) (Cassar 2005; Sabbioni et al. 2006). For example, concern has been expressed that an increase in biodeterioration of stone in Scotland can be expected due to higher temperatures and rainfall (Duthie et al. 2008). And in central Europe, the yearly number of humidity fluctuations crossing the deliquescence point of sodium chloride (~75 percent RH) are projected to increase two- to four-fold by the end of the century due to drier summers, which is likely to increase damage from salt crystallization (Brimblecombe and Grossi 2007; Grossi et al. 2008). Climate change is a very real threat to our monuments and cannot be ignored.

Salts

Along with air pollution, soluble salts represent one of the most important causes of stone decay. Salts cause damage to stone in several ways. The most important is the growth of salt crystals within the pores of a stone, which can generate stresses that are sufficient to overcome the stone's tensile strength and turn the stone to a powder. The deterioration of many of the world's greatest monuments can be attributed to salts, from Angkor Wat (Siedel, von Plehwe-Leisen, and Leisen 2008) to Venice (Lazzarini et al. 2008), and from Petra (Simon, Shaer, and Kaiser 2006) to the Great Sphinx of Giza (Reed 2002).

There are many ways in which stonework can become contaminated with salts. Air pollution is a major source of sulfates and nitrates. Other sources include the soil, from which salts may be carried into masonry by rising damp; salts blown by the wind from the sea or the desert; deicing salt; unsuitable cleaning materials; incompatible building materials; garden fertilizers; and, in the case of some medieval buildings, the storage of salts for meat preservation or even for gunpowder.

The growth of damaging salt crystals is usually attributable to crystallization, caused by the evaporation or cooling of salt solutions within the stone. In the past, there was much reference to "hydration damage," building on the fact that some salts can exist in more than one hydration state. The prime example is sodium sulfate, one of the most damaging of soluble salts, which can exist as the anhydrous salt thenardite (Na_2SO_4) or the decahydrate mirabilite ($Na_2SO_4 \cdot 10H_2O$) (Doehne 2003; Espinosa Marzal and Scherer 2008a). Thenardite increases in volume by more than three times on conversion to mirabilite, and it has been argued that this growth in volume was the cause of so-called hydration damage. However, it is now recognized that a crystal cannot magically transform from one form to the other without first dissolving and then recrystallizing in the new form. Hydration damage thus becomes a special case of crystallization damage (Doehne 1994; Flatt and Scherer 2002; Flatt 2006). Having said that, it is now becoming recognized that the sodium sulfate system presents yet further challenges, with researchers demonstrating the crystallization of the metastable heptahydrate ($Na_2SO_4 \cdot 7H_2O$) in preference to mirabilite in some circumstances (Hamilton, Hall, and Pel 2008; Saidov and Pel 2008).

Salt damage does not occur only in an outdoor environment, where the stone is subjected to cycles of rainfall and subsequent drying. It can also take place indoors, through the hygroscopic action of the salts. Severe damage to stonework held in uncontrolled museum environments is not uncommon (Hanna 1984; Rodríguez-Navarro et al. 1998).

On first sight, it appears surprising that salt damage should occur at all. Crystallization, for example, results in the formation of crystals that occupy a smaller volume than the solution from which they have been deposited. Is there not ample room for the crystals to develop in the pores, without the necessity of forcing the pore walls apart? However, this simplistic view overlooks the dynamic aspects of stone decay (Yu and Oguchi 2009). A stone may be fed constantly with salt-bearing moisture from the soil, for example, so that salts are constantly accumulating at the point of evaporation. Detailed analyses of this situation are given by Lewin (1982) and by Hall and Hoff (2007) and in a useful new book by the Italian engineer Edgardo Pinto Guerra, *Risanamento di murature umide e degradate* (Restoration of Damp and Deteriorated Masonry Walls) (2008). Work in Rhodes shows that the amount of salt is correlated to the severity of damage to the stone (Stefanis, Theoulakis, and Pilinis 2009).

Several tables of salt levels that are considered potentially hazardous for porous materials have been published in Germany (Wissenschaftlich-Technische-Arbeitsgemeinschaft für Bauwerkserhaltung und Denkmalpflege e.V. 1999), Austria (Österreichisches Normungsinstitut [ON] 2006), and France (Ministère de la culture et de la communication 2003). Simply measuring the concentration of salt in stone captures only part of the issue, since substrate characteristics (resistance to salt weathering) as well as the severity and frequency of environmental fluctuations are important in determining rates of salt damage (Doehne 2002). Any proposed international norm for salt levels in porous materials would have to take these factors into account, in addition to addressing the issue of identifying a method for measuring salt levels in building materials that is less costly than ion chromatography.

Modeling by Hall, Hoff, and Hamilton (2008) shows that in the UK rising damp can typically transport several hundred liters of moisture per year, per linear meter of stone, which can easily result in the accumulation of salts even from dilute groundwater solutions. The accumulation of salts and whether they crystallize on the surface or as a subflorescence has been related to the interfacial properties (wetting) and to the transport properties of the liquid. For example, De Witte (2001) and Miquel and others (2002) have clearly shown that subflorescence can develop at the interface between treated and untreated stone, and subsequent contour scaling can be due to the presence of water repellents. In lab experiments, Shahidzadeh and others (2008) have confirmed that interfacial properties are of key importance for where and how the crystals form. Pel, Sawdy, and Voroninaa (2010) have described the Peclet number[6] as a useful parameter that relates the rate of advection of a flow to its rate of diffusion in building materials. When advection dominates, salts will tend to accumulate at the surface of a stone. When diffusion dominates, ions will be more widely distributed.

There have been major advances in our understanding of salt weathering over the past fifteen years (Rodríguez-Navarro and Doehne 1999), although research into ways to prevent, mitigate, and treat the problems has lagged somewhat behind. The first big advance is related to the behavior of solutions containing more than one salt—the situation that is almost invariably found in practice. It is a straightforward process to predict the environmental conditions under which a single salt will pick up moisture from the air and subsequently lose it (causing damage by crystallization). However, the conditions under which salt mixtures will cause damage are much more difficult to predict and entails thermodynamic modeling. This work has advanced in several steps (Steiger and Zeunert 1996; Price 2000; Steiger 2006; Sawdy and Heritage 2007; De Clercq 2008; Franzen and Mirwald 2009). In an example that highlights the importance of understanding material behavior, recent work has shown that the type of salt is critical in determining if damage may occur. A pillar at Angkor Wat with severe erosion at its base was found to contain the same amount of salt in damaged and undamaged areas, leading to questions about whether salts were or were not the cause of the damage. Thermodynamic calculations subsequently showed that there were differences in the salt type present that explained the damage pattern, with highly hygroscopic salts that did not crystallize often being present in the undamaged zone and salts that crystallized frequently being present in the damaged zone (M. Steiger, personal communication). Computer programs can now predict the "safe" ranges of temperature and relative humidity in which crystallization damage may be minimized (Sawdy and Price 2005; Simon and Doehne 2006b; Price 2007; Steiger, Kiekbusch, and Nicolai 2008). Inevitably, there are limitations, the most important being that the programs can predict only what will happen under equilibrium conditions; they say nothing about the *rate* at which it will happen (Prokos and Bala'awi 2008).

The second important area of research is concerned with the mechanism by which damage occurs. Some of the papers are quite daunting, but the ideas are essentially quite simple (Hamilton and Hall 2004; Espinosa Marzal and Scherer 2009). Consider a crystal bridging a pore and exerting a pressure on the pore walls. If it is to grow any further, and thereby do damage, it is necessary for the surrounding solution to be able to get in between the crystal and the pore walls. If the pressure gets so high that this solution is squeezed out, no further growth can occur and there will be no damage. There is therefore a tug of war (or perhaps "push of war" would be more appropriate) between various opposing forces related to the surface energies of the respective stone/solution/crystal interfaces. As the surfaces of the salt crystal and the pore wall get to within 10 nm or so, repulsive forces arise due to the mismatched surface energy of the mineral surfaces and this keeps them apart, much like what happens when an attempt is made to push two magnets together. This mismatched surface energy can be thought of as the degree of lattice compatibility or incompatibility between two minerals. NMR, AFM, thermo-mechanical analysis (TMA), and environmental scanning electron microscopy (ESEM) have provided direct evidence of the existence of salt crystallization or disjoining pressure (Rijniers et al. 2005; Hamilton, Koutsos, and Hall forthcoming; Balboni et al. forthcoming) and the process has been modeled as well (Espinosa, Franke, and Deckelmann 2008). Future work on the calculation and measurement of the actual supersaturation that occurs in a porous medium at the exact moment of salt crystallization will help in understanding the widespread variability in resistance of various building materials to salt weathering. Research is continuing using synchrotron X-rays, NMR

(Hamilton, Hall, and Pel 2008), and differential scanning calorimetry (DSC) (Espinosa Marzal and Scherer forthcoming). Some additional work on direct measurement of the disjoining pressure using AFM is also needed to help clarify some of these issues.

The size of the substrate pores is important in salt weathering, as shown by measurements and models developed over the past decade (Scherer 1999, 2000; Steiger 2005a, 2005b). Under equilibrium conditions, a crystallization pressure can only occur in the smallest pores (less than 30 nm) (Rijniers et al. 2005). Since most types of stone have few pores in this range, it is predicted that most salt weathering damage takes place during nonequilibrium conditions, such as rapid drying. Another way that damage increases is when the stone pores fill with salt, which modifies the pore size distribution, essentially creating small pores where crystallization pressure can occur, even under equilibrium conditions. This may help explain the sudden onset of some salt weathering problems, since damage may not start until the pores are full of salt.

Other researchers have looked at the initial nucleation and growth stages of crystals in pores, with a view to inhibiting nucleation or modifying the shape and size of the crystals that form by using trace amounts of the chemicals commonly used to inhibit mineral scaling on pipes in industrial applications (Selwitz and Doehne 2002; Rodríguez-Navarro, Hernandez, and Sebastian 2006; Cassar et al. 2008; Ruiz-Agudo, Putnis, and Rodríguez-Navarro 2008). In some ways it is a risky strategy, for it has long been known that crystallization pressure is related to the degree of supersaturation of the solution from which the crystals grow. If nucleation is inhibited or postponed, this will lead to even higher levels of supersaturation, so that damage (if and when it does occur) may be more severe than it might have been. Modifiers may also behave differently when in solution than when absorbed to stone surfaces.

A further aspect of recent research concerns the role of wind in salt weathering or alveolar weathering.[7] The formation of alveolar or honeycomb weathering patterns is apparently due to the preferential accumulation of salt in sheltered hollows, where it was not washed away by rain as it would be in the ridges surrounding the hollows, and the protective effects of endolithic microbes (Laue et al. 2005; Siedel 2008; Mustoe 2010). The hollows are the last place to dry, and thus the place where salts tend to accumulate. Also, the ridges would generally be dry, while fluctuations in moisture in the depth of the cavity would result in cycles of salt crystallization and further erosion at the deepest point of the cavity. The main source of salts is thought to be wind-borne dust from nearby playas or sea-salt aerosol (Kirchner 1996).

Early laboratory work on wind's effect on stone showed that the boundary between air-filled pores and solution-filled pores in a stone could be moved into the sample by placing a fan facing the stone (Mossotti and Castanier 1990). Thus, the location of salts (efflorescence or subflorescence) is due in part to the rate of air exchange at the surface of the material, and windy conditions can result in the crystallization of salts as a more damaging subflorescence, rather than as a surface efflorescence. More recent modeling indicates that the development of a uniform erosion pattern or a honeycomb pattern of weathering may be explained by differences in the duration of drying periods (Huinink, Pel, and Kopinga 2004). The researchers found that short drying periods tend to result in the accumulation of salts on the surface (resulting in more uniform erosion). For longer drying periods (slow evaporation rates), salts accumulate in sheltered areas with lower evaporation rates (tending to expand pits and resulting in honeycomb patterns). Experimental laboratory work has also shown that wind (Selwitz and Doehne 2002) and related rapid drying (Lombardo, Doehne, and Simon 2004) increases damage rates due to increases in salt supersaturation, and that variable weathering rates related to wind can result in honeycomb patterns (Rodríguez-Navarro, Doehne, and Sebastian 1999). Recent modeling of the effect of wind on the Sphinx found that areas of rapid erosion correlated with areas of high wind friction and enhanced drying (left shoulder and the top of the haunches) (Hawass 1998; Hussein and El-Shishiny 2009). Lab experiments and work at sites such as Petra in Jordan show that wind speed strongly influences the rate of damage and pattern of salt distribution (Bala'awi 2008). Pore blocking by salts also appears to be an important factor in controlling the pattern of salt weathering damage (Espinosa Marzal and Scherer 2008b; McCabe, McKinley, and Smith 2008; Espinosa Marzal and Scherer forthcoming) and may result in greater crystallization of salts as subflorescence.

Is crystallization the only way in which salts can cause damage? It seems not. It appears that they can also cause damage through stress from differential thermal expansion, since sodium chloride, for example, expands at about five times the rate of calcite at surface temperatures (Nocita 1987; Holmer 1998; Smith et al. 2005). Schaffer (1932) attributed this idea to Scott Russell. Salts also have a role to play in the weathering of stones that contain clay minerals (Snethlage and Wendler 1997; Rodríguez-Navarro et al. 1998; Scherer 2006; Scherer and Jiménez-González 2008), in some cases enhancing the swelling potential of these stones. While most of the damage from salts is physical, work shows that salt solutions enhance the dissolution of calcite (Ruiz-Agudo, Martín-Ramos, and Rodríguez-Navarro 2007) and the alteration of biotite (Silva and Simão 2009), quartz (Young 1987), and feldspars (Bernabe, Bromblet, and Robert 1995). And while one might expect salt weathering to have little in common with biodeterioration, recent work has found that halophilic bacteria are often present and may enhance physical damage mechanisms (Laiz et al. 2000; Papida, Murphy, and May 2000).

There are several recent reviews that give further details of research in this area. They include a thoughtful overview of the role of salts in the deterioration of porous materials by Charola (2000) and an excellent discussion of salts and crusts by Steiger (2003). Doehne (2002) reviews the scope and interdisciplinary nature of salt weathering in a paper that brings in perspectives from conservators, geomorphologists, and cement specialists. Simon and Doehne (2006a; 2006b) summarize a series of discussions and expert papers on salt weathering and masonry desalination. A special issue of the journal *Environmental Geology* was devoted to salt decay with three groups of papers devoted to salt weathering tests, salt behavior, and field studies (Steiger and Siegesmund 2007). A detailed summary of the fundamental basis of salt decay mechanisms can be found in Scherer (2004). "Salt Weathering on Buildings and Stone Sculpture" conferences were held in Ghent in May 2007 and in Copenhagen in October 2008 (Albertsen 2008); the next conference will take place in Cyprus in October 2011. Finally, a recent review of salt weathering calls for new field research on building material behavior and soluble salts (Gulotta et al. 2008).

Closely related to the issue of salt damage is the issue of damage from frost. The topic has been reviewed by Scherer and Valenza (2005) and Matsuoka and Murton (2008). In France, the standard on frost resistance of natural stone (Norm XP B 10-601, see LERM 2006) gathers all the tests to be performed and gives the appropriate thresholds, according to the destination of the stone in the building and according to local climate. Created in 1984, the standard is regularly revised to fit with field observations and climate change.

Inevitably, further questions remain. Why are certain types of stone much more vulnerable than other types to salt damage? Why are certain salts much more damaging than others? Is damage caused mostly by relatively rare environmental events (rapid cooling, drying, or condensation) or cumulative everyday stresses (humidity cycling)? What are the long-term effects of various conservation treatments, such as desalination or consolidation, on salt damage? How can desalination and preventive conservation efforts be enhanced? Can general agreement be achieved regarding the fundamental mechanisms of salt weathering? Can the salt damage process and weathering forms such as tafoni be accurately modeled using existing knowledge? How does the hydration of salts progress, and how are crystallization pressures sustained in situ? And, above all, how can the great fundamental strides of recent years be converted to practicable applications?

Biodeterioration

In 1932, in his classic report *The Weathering of Natural Building Stones*, Schaffer wrote:

> Living organisms also contribute to the decay of stone and similar materials and, although their action is, generally, of somewhat less importance than certain of the other deleterious agencies which have been considered, their study presents numerous features of interest. The effect of certain organisms, such as bacteria, is still a matter of controversy, but the effect of others, such as the growth of ivy, is generally considered to be detrimental.

In two of these areas, he is still remarkably up to date. There is controversy over the role of bacteria, and we still need to weigh the importance of biodeterioration against the importance of other causes of decay. However, recent work on ivy suggests that the shade and thermal stability provided by ivy on stone walls may be beneficial in certain situations (H. Viles, personal communication; see also: http://www.srs.ac.uk/scienceandheritage/presentations/Ivy_presentation2.pdf).

Biological growths on stone are both a blessing and a blight. Colorful lichens and creepers, such as ivy, can contribute an air of age and romance to a monument, and their removal can leave the stone looking stark and denuded. Nevertheless, many organisms contribute to the deterioration of stone, and it is necessary to find the right balance between appearance and longevity. The discussion surrounding this topic has become more complex and nuanced, as evidence has accumulated that complex biofilms in some situations may help to stabilize fragile stone surfaces and in other cases may strongly accelerate decay (Uchida et al. 2000; Chiari and Cossio 2004; Caneva et al. 2005; De Muynck, De Belie, and Verstraete 2010). For example, in laboratory experiments, biofilms have been shown to result in a 40–70 percent decline in dissolution rates of calcite (Davis and Lüttge 2005). In more recent work, the contribution of bacteria to dissolution or protection has been related to the amount and type of "food for microbes" present, such as nitrate versus ammonium ions and organic carbon species (Jacobson and Wu 2009). Research on the action of biofilms on silicate stones (granite and basalt) has shown they may enhance dissolution rates in some situations (Wu et al. 2007; Wu, Jacobson, and Hausner 2008). While additional work is needed, research in this area suggests that some surface patinas may be an effective natural protection for carbonate stones, while other biofilms, particularly in polluted environments, may be deleterious.

Bioremediation and biocides are related topics of recent research that are discussed later in the section on surface treatments in chapter 2.

The biological degradation of rocks is well known and has been studied for a long time: it is one of the weathering mechanisms responsible for the formation of soil. The deterioration of stone in buildings and monuments through the action of biological organisms has also been acknowledged since the mid-1960s, but the topic has received increasing attention over the past decade. Some of the literature is concerned primarily with the influence of organisms on the appearance of stone surfaces, while other research deals primarily with the deterioration of the stone itself. In the past, microbiologists studying this topic have tended to focus more on characterizing the species and ecosystems found on stone and less on the nature of the effects of biological agents on stone decay. This is changing with new research on how biofilms change the thermal, hygric, and mechanical behavior of stone, thus enhancing decay by increasing the duration of surface wetting and providing a source of organic acids and complexing agents.

Excellent reviews of the topic are provided by Warscheid and Braams (2000); Caneva, Gasperini, and Salvadori (2008); Warscheid (2008); and Scheerer, Ortega-Morales, and Gaylarde (2009). Other useful overviews are given by Wakefield and Jones (1998); Ciferri, Tiano, and Mastromei (2000); and Crispim and Gaylarde (2005). A burst of earlier reviews can be found in Gómez-Alarcon and de la Torre (1994); Jain, Mishra, and Singh (1994); May et al. (1993); and Tiano (1994). Krumbein and Urzï (1992) have set out a comprehensive terminology for describing aspects of biodeterioration on stone.

Much of the recent research has been centered on algae, lichens, and bacteria. Adamo and Violante (2000); Jie Chen, Blume, and Beyer (2000); Schiavon (2002); Wilson (2004); St. Clair and Seaward (2004); and Piervittori, Salvadori, and Isocrono (2004) have reviewed the action of lichens, confirming that their effects are both physical and chemical. Mechanical damage is caused by penetration of the hyphae into the stone and by the expansion and contraction of the thallus (the vegetative part of the fungus) under changes of humidity. Chemical damage, however, is more important and may arise in three ways: by the secretion of oxalic acid, by the generation of carbonic acid, and by the generation of other acids capable of chelating ions such as calcium. Field examples of damage from lichens to stone monuments have recently been described in contexts ranging from Persepolis to the Alhambra palace and the Jeronimos Monastery (Mohammadi and Krumbein 2008; Sarró et al. 2006; Ascaso et al. 2002).

The secretion of oxalic acid, which reacts with a calcareous stone to produce calcium oxalate, is of particular interest. A number of authors have noted the presence of calcium oxalate on the surface of stone monuments, where it can form part of a coherent and seemingly protective layer known as *scialbatura*. Del Monte and Sabbioni (1987), for example, have argued that *scialbatura* is caused solely by lichen activity, whereas Lazzarini and Salvadori (1989) have enumerated other possible causes, including the deliberate application of a protective coating. Correlating the environmental limits for lichen growth with the distribution of oxalate on Trajan's Column, Caneva (1993) found that the oxalate distribution pattern was the opposite of that expected for lichens and thus lichens were perhaps not the best explanation for the column's patina. Analysis of rock outcrops suggests that some oxalate patinas may be relics of past paleo-environments that were more suitable for lichen growth during an interval of greater surface moisture (Moore et al. 2000). Subsequent experimental work has shown that the alteration of an organic coating can result in the formation of calcium oxalate (Cariati et al. 2000). Analysis of oxalate patinas in the field has also provided support to the idea that microbial alteration of organic material contributes to oxalate formation (Casoli and Palla 2002). In addition, Monte (2003) has performed experiments showing oxalate can be produced by the action of fungi alone on marble. More on biomineralization can be found in chapter 2.

Researchers have found a range of microbes that are endolithic—colonizing the interior of porous stone (Walker and Pace 2007). There they take advantage of the light, moisture, and shelter found inside the stone (Caneva, Gasperini, and Salvadori 2008) and may also modify their surroundings (McNamara et al. 2006). It seems clear that the presence of endolithic communities should be assumed in many environments and stone types. Further study is needed of their role in stone alteration, their modification of moisture transport, and their interaction with conservation treatments such as consolidants and biocides.

The heavyweight controversy is saved for bacteria. They have long been implicated in stone decay, but acceptance of their role has sometimes been hindered by the emotive stance of some researchers, who have appeared determined to see bacteria and nothing else. A troubling number of authors have noted high numbers of bacteria in decaying stone, in comparison to low numbers in sound stone, and have concluded that the bacteria cause the decay. However, an alternative explanation could be that decayed stone presents a preferred habitat for the bacteria.

Bacteria that attack stone chemically fall into two groups: autotrophic bacteria derive their carbon from carbon dioxide (CO_2), and may derive their energy from light (photolithotrophs) or from chemical redox reactions (chemolithotrophs). Heterotrophic bacteria, by contrast, utilize organic compounds on the stone to derive their carbon. Autotrophic bacteria include those that are capable of oxidizing sulfur and nitrogen compounds to produce sulfuric acid and nitric acid, respectively. They are one more means, therefore, by which air pollutants such as sulfur dioxide and nitrogen oxide are turned into sulfates and nitrates. This underlines the difficulty of separating out the individual causes of stone decay; several different factors may play integral roles in the overall decay process. The question remains of whether bacteria or catalyzing metal compounds, for example, are the main routes of sulfate production. However, if oxidation by both bacteria and metal compounds is rapid by comparison with the rate at which sulfur dioxide arrives at the stone surface, then the arrival rate will be the rate-determining step, not the route taken. The synergistic effects of air pollution and biofilm formation have been researched, with the finding that there is strong evidence that biofilms enhance the absorption of air pollutants (Young 1996; Mansch and Bock 1998).

Heterotrophic bacteria produce chelating agents and organic acids that are weaker than the inorganic acids produced by the sulfur-oxidizing and nitrifying bacteria. They have received comparatively little attention, but their role in deterioration is well established nonetheless (Lewis, May, and Bravery 1988; Saarela et al. 2004; Gorbushina 2007; Maruthamuthu et al. 2008).

Recent work on bacteria has helped to quantify the effect they have on the dissolution of limestone, with one example showing a two-fold increase in the laboratory dissolution rate compared to control limestone samples (McNamara et al. 2005). Bacteria typically produce slime or extracellular polymeric substances (EPS) as part of a complex biofilm made up of polysaccharides, water, and proteins that has been shown to change the dissolution rate and dissolution pit morphology on samples of limestone (Perry et al. 2004). Damage from bacteria in the field has been described by McNamara and others (2006) and Mansch and Bock (1998).

Biodeterioration studies of important cave painting sites such as Altamira have resulted in recent advances in understanding. Researchers have found that cyanobacteria and algae (phototrophs) and networks of heterotrophic bacteria increase stone deterioration through their metabolic products, biomediated dissolution, and mechanical alteration, such as scaling (Cañaveras et al. 2001). As expected, the control of moisture, food, and light levels appears to be the most effective prevention method (Dornieden, Gorbushina, and Krumbein 2000; Zammit et al. 2008).

Environmental control of cave environments has proven to be complex and controversial, as in the cave at Lascaux, where efforts to limit the introduction of fungal strains through the use of a formaldehyde foot-wash treatment for visitors resulted in the growth of a formaldehyde-resistant strain of white *Fusarium solani* fungus (Bastian et al. 2007; Dupont et al. 2007; Jurado et al. 2009; Bastian and Alabouvette 2009; Bastian et al. 2009; Bastian, Alabouvette, and Saiz-Jimenez 2009). This condition may have been exacerbated by the installation of a new ventilation system (Brunet, Malaurent, and Lastennet 2006; Lacanette et al. 2009). Computer modeling of the airflow at the Lascaux cave suggests that reducing the airflow may help avoid future damage (Malaurent et al. 2007).

The role of halophilic microbes (mostly archaea, with some bacteria) is important in stone decay (Laiz et al. 2000; Saiz-Jimenez and Laiz 2000). A significant and open question is if hydroscopic salts may raise moisture levels to the point where halophilic microbes increase in abundance, setting the stage for further microbial development of adjacent areas of stone.

Differential Stress

While air pollution, salts, and biodeterioration capture the lion's share of attention, there are advances in our understanding of other, often related decay mechanisms that are worth some consideration. Reviewing the recent literature on stone conservation, it is clear that there is an important trend in decay mechanism research that is focusing on what is called here (for want of a better term) "differential stress." This decay mechanism includes the effects of wet/dry cycling, clay swelling, differential hygric stress, differential thermal stress, and stress from differential expansion rates of material in pores (such as salts or organic material) versus in the stone. The general idea is that treatments, salts, water films, or biofilms—anything that causes the stone surface to react differently than the interior—can result in a shear stress, crack propagation, and, eventually, surface parallel detachment (e.g., flaking). For example, significant shear stress is generated when, during a brief afternoon rain, the surface of a clay-containing stone swells, while the interior of the stone remains dry (Doehne et al. 2005). This would be considered an example of differential hygric stress and is typically found on the corners of stones such as Sydney sandstone and Portland brownstone. As mentioned earlier, sodium chloride expands at approximately five times the rate of calcite at surface temperatures, so decay in limestone from this mechanism would be an example of stress induced by differential thermal expansion (Nocita 1987; Holmer 1998; Smith et al. 2005). Note that salts naturally tend to accumulate near the stone surface, setting up differences in how the two parts of the stone (surface and interior) react to environmental changes. Modeling has shown that there may be a particular depth beneath the stone surface where moisture is present and salts may accumulate (Snethlage and Wendler 1997). This depth is often the same as the thickness of stone flakes or scales. Differential thermal expansion stresses may also be induced at the interface between minerals having different colors, for instance in granites exposed to direct sunlight (Casta 1988). Field measurements of stone surfaces show that rapid thermal variations are more common than previously thought (Molaro and McKay 2010).

Organic material in pores, whether it is a polymeric consolidant or originating from biological sources, may expand significantly faster than the stone on wetting (Laurenzi Tabasso 1995; GCI and IHAH 2006). Work by Yang, Scherer, and Wheeler (1998) highlighted the importance of making sure consolidants have thermal expansion properties similar to the substrate being treated. Some recent work on thermal damage to stone reveals that it is an important decay factor, and stress may result from differential heating, such as when areas of stone undergo short-term cooling events from shade (Weiss et al. 2004; Gómez-Heras, Smith, and Fort 2006; Hall and André 2006; Sumner, Hedding, and Meiklejohn 2007; Gómez-Heras, Smith, and Fort 2008). This effect may be more pronounced at high-altitude sites such as Tiwanaku in Bolivia (Maekawa, Lambert, and Meyer 1995), where the drop in temperature when a cloud blocks the sun is substantial.

Work by Warke and Smith (1998) found that climate chamber simulation studies do not take into account the important effects of radiant heating and thus are not representative of field conditions. The balance between thermal and hygric damage is addressed in work at Petra by Paradise (2002; 2005). Research on clay swelling has advanced significantly based on the research at Princeton University (Wangler and Scherer 2008; Duffus, Wangler, and Scherer 2008; Jiménez-González, Rodríguez-Navarro, and Scherer 2008), where it was found that shear forces can cause buckling of wetted stone surfaces and that intracrystalline swelling of clay is the primary mode of swelling for Portland brownstone, despite the proportion of swellable clay being only 1 percent of the stone. The clay is present as a cement at sand grain boundaries, permitting the clay sufficient leverage in brownstone. Osmotic swelling (salt-activated clay swelling) was found to be important in sepiolite-rich Egyptian limestone (Rodríguez-Navarro et al. 1998). Understanding the relative role and dynamics of differential stress as it relates to air pollution, biodeterioration, salt weathering, and conservation treatments remains an area for future research.

Intrinsic Problems

"Intrinsic problems" (or "inherent vice") is an expression that places the blame for stone decay squarely on the material, rather than the particular environment. Every region seems to have a stone that ought to have remained in the ground, rather than being used to create sculpture, monuments, and buildings. Some examples include Reigate stone in the UK (which contains unstable silica, glauconite, and smectite), Lausanne molasse in Switzerland (containing swelling clays), and the Lecce limestone of Italy (with a porosity of 50 percent). These "difficult stones" appear to account for a disproportionate share of stone conservators' attention. Recent research into these problematic stones includes several studies of Lecce and similar highly porous limestones (Fratini et al. 1990; Calia et al. 2004; Atzeni, Sanna, and Spanu 2006).

As far as Reigate stone is concerned, decay of this material is not a new phenomenon. Reporting in 1713 on the condition of Westminster Abbey, Sir Christopher Wren wrote, "the Ashlar of the whole fabric . . . is disfigured in the highest degree . . . and the stone is decayed 4 inches deep and falls off perpetually in great scales." He comments wryly that Reigate "is a stone that would saw and work like wood, but not durable, as is manifest" (as quoted in Prudon 1975). An alternative point of view notes that the decay of Reigate stone is mainly confined to surface layers and has not been responsible for structural failure (Lockwood 1994).

Work on Swiss molasse and similar stones has included the use of grouts for the extensive detached areas often found on buildings and a new treatment for reducing the swelling of clays (discussed in more detail in chapter 3) (Jiménez-González and Scherer 2004; Rousset et al. 2005).

One intrinsic issue that researchers have puzzled over for several decades is the bowing of thin marble slabs on emblematic modern buildings, such as the Amoco building in Chicago, the Grande Arch de la Defénse in Paris, and Alvar Aalto's Finlandia city hall in Helsinki. Substantial recent research has found that differential expansion of calcite enhanced by moisture, microstructure, and differential residual strains in the marble is the main cause of this problematic and still somewhat mysterious phenomena (Siegesmund, Koch, and Ruedrich 2007; Grelk et al. 2007; Siegesmund, Ruedrich, and Koch 2008; Malaga, Schouenborg, and Grelk 2008; Marini and Bellopede 2009).

Notes

1 System dynamics deals with understanding the behavior of complex systems over time. It is an approach that uses internal feedback loops and time delays to characterize the entire system and nonlinear behaviors.

2 Major programs were coordinated by the European Union through its STEP and Environment initiatives, the NATO Committee for the Challenges of Modern Society, the United Nations Economic Commission for Europe (UNECE), the US National Acid Precipitation Assessment Program (NAPAP), the UK National Materials Exposure Programme, and the German Bundesministerium für Forschung und Technologie (BMFT).

3 EC project "Carbon content and origin of damage layers in European monuments–CARAMEL" (EVK4-CT-2000-00029). Related EC projects on air pollution and climate change include: MULTI-ASSESS (2002–5), CULTSTRAT (2004–7), and Noah's Ark (2004–7).

4 Sulfur dioxide is detectable to the human nose at concentrations of about 0.5–0.8 parts per million (1400–2240 $\mu g/m^3$).

5 SO_2 limits: WHO (2008): 20 $\mu g/m^3$; US EPA (1997): 80 $\mu g/m^3$; EU (2008): 20 $\mu g/m^3$

6 The Peclet number is the ratio of the rate of solute transport by advection to the rate of transport by molecular diffusion. For $Pe \ll 1$, diffusion dominates and ion transport proceeds according to the concentration gradient. If $Pe \gg 1$, advection dominates and ion transport takes place due to capillary water flow. The Peclet number is defined at the macroscopic scale of the bulk porous material.

7 "Alveolization is a kind of differential weathering possibly due to inhomogeneities in physical or chemical properties of the stone. Alveolization may occur with other degradation patterns such as granular disintegration and/or scaling. In arid climates large alveoles of meter size are frequently formed (e.g., Petra Jordan)" (Vergès-Belmin 2008, 28).

Chapter 2

Putting It Right: Preventive and Remedial Treatments

When confronted with decaying stonework, one's immediate instinct is to "do something about it." Traditionally, this has meant doing something to the stone: perhaps patching it up with mortar, applying some kind of protective coating, or cutting out decayed stone and replacing it with new stone. Regular maintenance is vitally important, wherever practicable; William Morris (1877) wrote of the need to "stave off decay by daily care," and in a textbook for conservators encouragingly titled *Preventive Conservation of Stone Historical Objects,* Domaslowski (2003) persuasively argues that routine maintenance is an often-underappreciated aspect of preventive conservation. Now, however, there is an increasing emphasis on doing something not only to the stone itself but also to the environment in which the stone is found. This reflects a growing awareness of the importance of preventive conservation, of the principle of minimum intervention, and of the need to limit the use of materials that might prove harmful to either the stone or to the environment. Also, now that there is a better understanding of decay mechanisms, a conservation strategy can be designed to reduce the rate of damage by focusing on points of leverage that can mitigate some decay processes. An interesting example is the use of multispectral satellite images of a historic city to provide an automated assessment of the condition of the roofs, where building degradation often begins (Gonçalves et al. 2009).

PREVENTIVE CONSERVATION

Doing something to the stone's environment is not simply a matter of temperature and relative humidity. Preventing damage can embrace a very wide range of topics: legislation to protect individual buildings and monuments, pollution control, traffic control, control of groundwater, visitor management, and disaster planning (Baer 1991; Baer and Snethlage 1997; Baer and Snickars 2001). Such topics may seem remote from the problems of an individual block of stone, but they are nonetheless of great importance. Other areas of preventive research on immovable stone heritage have included shelters, wind fences, and reburial (Demas 2004; Teutonico 2004), as well as modeling of interior environments to help determine needed interventions (Albero et al. 2004).

Preventive conservation measures of more immediate effect are usually concerned with keeping water out of the stone and with controlling the relative humidity and temperature of the air around the stone. This is relatively easy for stone artifacts within a museum, and it may also be feasible for stone masonry that is exposed on the interior of a building (Price and Brimblecombe 1994; Price 2007). It is less easy for stonework on the outside of a building, although a dramatic example of this approach is provided by the glass envelope constructed over the ruins of Hamar Cathedral in Norway.

More modest protective shelters are frequently used on the outside of a building to protect those features that are particularly important. They may be part of the original design (for example, a canopy protecting a statue in a niche), or they may be a later addition. As an extreme measure, they may enclose the feature altogether. Their purpose is to reduce the amount of rain that reaches the stone and, insofar as is practicable, to stabilize the temperature and moisture content of the stone. If the shelter is a later addition, it is likely to be visually intrusive—unless it is so small as to serve little purpose.

Few studies have been undertaken of the design requirements of such shelters, and it is possible that their benefits are more psychological than actual. This has been evaluated in practice in only a few cases (Agnew et al. 1996; Aslan 2007). One case study is at the site of the Hieroglyphic Stairway, in Copán, Honduras, where a simple canvas shelter has prevented lichen growth and the swelling of the clay-containing stone due to frequent rainstorms (Doehne et al. 2005; GCI and IHAH 2006). A second case study, which calculated protective indices for several styles of shelter at the archaeological site of Joya de Ceren in El Salvador, found that evaporation was reduced and thermal and relative humidity stability improved in several cases (Maekawa 2006). A further useful study was undertaken for a pavilion at Chartwell, Sir Winston Churchill's country house in Kent, England (Lithgow, Curteis, and Bullock 2007). The pavilion was open on two sides, and interior decoration suffered from condensation events, which were mitigated by roof repairs and a temporary wall to buffer the microenvironment during winter.

The main purpose of relative humidity control or buffering is to reduce damage from salt and moisture cycles. The humidity regime required to prevent damage in a stone or a wall painting that is contaminated with a single salt is well established. However, stone is more commonly contaminated with a mixture of salts. As discussed in the section on salts in chapter 1, the behavior of salt mixtures is complex (Steiger and Zeunert 1996; Price 2000; Steiger 2005b; Sawdy and Heritage 2007; De Clercq 2008; Franzen and Mirwald 2009), and there are now methodologies to help with selecting appropriate humidity ranges, even for complex mixtures (Bionda 2004). Arnold has proposed a methodology for reducing salt damage to wall paintings by monitoring the relative humidity and temperature, and observing salt efflorescence over the course of one year (Arnold and Zehnder 1991; Arnold 1996). Then, the periods where salts appear can be correlated with the environmental parameters and the environmental conditions modified to reduce the incidence of salt crystallization events (Laue, Bläuer Böhm, and Jeannette 1996). There is increasing evidence that drying rates are important and that even a small reduction in drying rate can result in salts crystallizing on the surface as relatively harmless efflorescence (Selwitz and Doehne 2002). This was the logic behind the suggestion that a row of trees be planted to help protect salt-laden structures at the site of Port Arthur in Australia (Thorn and Piper 1996).

The remainder of this chapter is devoted to research related to active conservation: doing something directly to the stone itself. In keeping with the title of this volume, this chapter is not a handbook of repair techniques. Information on the routine practice of stone conservation is available elsewhere (Ashurst and Ashurst 1988; Ashurst and Dimes 1998; Ashurst 2007; Snethlage 2008).

ACTIVE CONSERVATION: CLEANING

Cleaning is often one of the first steps to be undertaken after a condition survey has been completed. As expected, carbonate materials are the most reactive to acidic pollution and thus have received the lion's share of attention in studies of stone cleaning. By removing the dirt, one can better see the condition of the underlying stone and thus judge what further conservation may be necessary. Cleaning may also serve in some circumstances to remove harmful materials from the surface. However, the primary reason for cleaning will often be the dramatic change in appearance that can be achieved. A dirty building or monument does not look well cared for, and the dirt may well obscure both fine detail and major architectural features. Nonetheless, there are those who would argue that cleaning contravenes one of the fundamental principles of conservation—reversibility—and that by removing the dirt one is removing both the sense and the evidence of history.

From a morphological point of view, the original stone surface may be present under a layer of soot or black crust. However, the stone cannot be considered original from the chemical point of view, having undergone a series of changes as the surface equilibrates with its varying environment (Vergès-Belmin 1994; Smith, Gómez-Heras, and McCabe 2008). Different types of gypsum crust morphology have been used as criteria for determining the appropriate degree of gypsum removal, and in some cases it has been deemed no longer a desirable goal to eliminate all gypsum from stone surfaces (Bromblet and Vergès-Belmin 1996; Siegesmund et al. 2007). After removal of black crusts, the persistence of a gypsum layer bearing no airborne particles may indicate that the original surface has been preserved. This type of layer is approximately 30–500 µm thick. It cannot be recognized with the naked eye; however, it is often detected in cross sections using optical microscopy, ESEM, or EDS (energy dispersive X-ray spectrometry) (Vergès-Belmin 1994). In other cases, a clear gypsum layer occurs underneath the fragile, hardened stone surface and therefore, when reached, it means that the original surface is completely gone (José Delgado Rodrigues, personal communication).

A wide range of techniques is available for cleaning stone, ranging from those that are intended for use on large facades to those that are intended for meticulous use on finely carved and delicate sculpture. Techniques are reviewed by a range of researchers and practitioners: Fassina 1994; Andrew, Young, and Tonge 1994; Ashurst 1994; Cooper, Emmony, and Larson 1995; BSI 2000; Vergès-Belmin and Bromblet 2000; Rodríguez-Navarro et al. 2003; Normandin et al. 2005; Worth 2007. This is an area where much progress has been made in the past twenty years, although only a portion is reported directly in the literature. The basic techniques have remained largely the same, although they have become more refined. This reflects an increasing awareness of the damage (and consequent litigation) that may be caused by inappropriate or overenthusiastic cleaning and also of the environmental issues posed by the use of certain chemicals or excessive quantities of water (Maxwell 1996; Young, Urquhart, and Laing 2003). With some exceptions, such as latex cleaning films, developments have largely come about through care and attention on-site rather than in the laboratory. These lessons from the field have been consolidated into guidelines (BSI 2000; Young et al. 2003).

It should be noted that any cleaning method requires judgment and an agreed-upon definition of the target cleaning level before the work begins. For example, in the present urban environment, uncleaned limestone surfaces may range in color from white (where water runoff has taken place) to dark brown and black, depending on the amount of accumulated dirt. All of these surfaces differ substantially from the "original" freshly cut surfaces, and establishing a target level of cleaning is not an easy task when a single building may contain a wide range of surfaces.

A number of authors have emphasized the damage that can be caused by cleaning: loss of surface, staining, deposition of soluble salts, or making the stone more vulnerable to pollutants or biological growths. They include Maxwell (1992); MacDonald, Thomson, and Tonge (1992); Young and Urquhart (1992); Andrew, Young, and Tonge (1994); Maxwell (2007); and Delegou and others (2008). It is undoubtedly the case that very severe damage can arise, but a degree of skepticism would perhaps be justified over "damage" that is observable only through a scanning electron microscope.

In most cleaning methods no attempt is made to collect the dirt and detritus, which is instead allowed to run down the stone and pass into the drains. Some attention is now being given to techniques that collect the detritus and, for example, permit recycling of the abrasive (Hoffmann and Heuser 1993). A commercial system has been developed that uses fine powders and an air extraction system to capture the debris. This and similar methods have seen wide application (Vergès-Belmin and Bromblet 2000; Iglesias, Prada, and Guasch 2008).

The effectiveness of a cleaning technique is usually assessed subjectively, although objective procedures have been described by many authors (Werner 1991; Young 1993; Andrew, Young, and Tonge 1994; D'Urbano et al. 1994; Vergès-Belmin 1996a; Kapsalas et al. 2007; Hauff,

Kozub, and D'ham 2008). Vergès-Belmin (1996b) gives a particularly useful overview of methods for evaluating cleaning treatments for stone. Recent work has shown that quantitative measurements of color change after stone cleaning vary considerably, mainly due to the action of hygroscopic salts (Vergès-Belmin, Rolland, and Leroux 2008). Precautions should be taken to account for the influence of salts when making such measurements. When discussing color change due to cleaning, it should be made clear that once aged, the stone surface can never be returned to the freshly cut color. Color can be used as criteria for cleaning only when a "reference surface" is defined and taken as a target for the cleaning level to be reached in the intervention. Color changes related to laser cleaning are dealt with in the next section.

Laser Cleaning

Using lasers to clean stone is now routine, and large-scale commercial application of laser cleaning has become more common over the past fifteen years (Dajnowski, Jenkins, and Lins 2009). Its great attraction is that it does not entail any physical contact with the stone and so lends itself to the cleaning of very delicate surfaces. There are no solvents or water to redistribute potentially harmful salts. The technique is selective and sensitive in terms of the degree and control of removal. The principle is essentially simple: a laser beam impacts the surface, and the energy of the infrared beam is dissipated by the sudden heating and expansion of light-absorbing material on the surface, such as particles rich in carbon, and the nearly instantaneous vaporization of moisture in the surface layer, which acts to remove surface dirt. Spraying the surface with water just before laser cleaning can enhance the effectiveness of the treatment (Siedel, Neumeister, and Sobott 2003). For light-colored stones with dark surface deposits, the infrared beam continues to be absorbed while the stone remains soiled and cleaning proceeds. Once the dirt has been removed, however, the light is reflected by the clean surface, and no more material is removed. This is not the case for biotite-bearing granites and painted stones, where laser cleaning may not be appropriate. The technique is described in detail by a number of authors, including Cooper, Emmony, and Larson (1993); Cooper (1998); Maravelaki-Kalaitzaki, Zafiropulos, and Fotakis (1999); and Orial and others (Orial and Riboulet 1993; Orial, Vieweger, and Loubiere 2003).

With early systems, the speed of cleaning was comparable to that achieved with a pencil-sized air-abrasive gun. The use of optic fibers to transmit the laser beam was a significant advance (EC project: LAMA—LAser MAnuportable pour le nettoyage des façades courantes et des monuments historiques; BRITE/EURAM BRE CT93-560). Now entire facades have been laser cleaned (Pini, Siano, and Salimbeni 2000), including the town hall in Rotterdam (Nijland and Wijffels 2003), and many monuments in Poland (Koss and Marczak 2008). The technique is seeing additional testing and application in the United States as well (Normandin et al. 2007).

Current research is aimed at selecting the optimal wavelength and pulse energy; at examining the effects on the stone, both physical and chemical; at comparing the performance of lasers with other cleaning techniques; and at identifying possible hazards to the operator (Vergès-Belmin et al. 2003; Bromblet, Labouré, and Orial 2003; Rodríguez-Navarro et al. 2003). The use of a laser requires special caution when cleaning surfaces with traces of polychromy (Fassina, Gaudini, and Cavaletti 2008). A set of conferences devoted to the use of lasers in art conservation (Lasers in the Conservation of Artworks, or LACONA) has taken place every two years since 1995: for example, Liverpool in 1997 and Madrid in 2007. A European Cooperation in Science and Technology project on the topic of artwork conservation by laser, funded by the European Science Foundation, ran from 2000 to 2006 and resulted in a handbook available for download (http://www.cost.esf.org/library/publications/05-40-Cleaning-Safely-with-a-Laser-in-Artwork-Conservation).

Further development of equipment has taken place, identifying, for example, the appropriate means and timing of delivering the laser pulse to the surface of the stone (Margheri et al. 2000; Mazzinghi and Margheri 2003; Dogariu et al. 2005; Siano et al. 2008). An important issue with laser cleaning is the color of the cleaned surface. In some cases, a yellow surface layer is revealed, which in some examples is related to previous restoration treatments (Vergès-Belmin and Dignard 2003; Zafiropulos et al. 2003; Gaviño et al. 2004; Gaviño et al. 2005; Vergès-Belmin and Laboure 2007; Andreotti et al. 2009). Color changes after laser cleaning may happen due to modifications in the substrate (pink feldspars, for instance), to modifications in any covering colors, or to changes in deposited dirt particles. The last situation may indicate that the target cleaning level has not been reached.

Latex Poultice Method

An important challenge for stone conservation has been the cleaning of large, public interiors, such as cathedrals, while allowing them to remain open during the process. This stricture generally rules out the use of toxic chemicals and abrasives. One innovative response to this challenge has been the development over the past fifteen years of the latex poultice method; it is known commercially as Arte Mundit. Originally developed as an improvement to the Mora poultice (Woolfitt and Abrey 2000) by Eddy De Witte (De Witte and Dupas 1992) as a spray-on film containing EDTA (ethylene diamine tetra acetic acid) and other additives, it has seen adaptation and application to a wide range of sites, including St. Paul's Cathedral in London (Miget 2000; Odgers 2003; Jacobs 2004; Stancliffe, De Witte, and De Witte 2005; Odgers 2006; Allanbrook and Normandin 2007). The method is best used on sound interior surfaces. If the soiling has been trapped in an encrustation such as a gypsum crust, the latex poultices no longer work. Recent research on latex poultices has raised the issue of residues left on stone surfaces by the method, which deserves further study (Morasset 2008; Morasset et al. 2009), and there may also be concern over unintentional mechanical damage to friable surfaces during removal. There are interesting parallels between the residue issue and the use of gels for the cleaning of paintings (Stulik et al. 2004).

Biological Cleaning

Hempel (1976) was one of the first to raise the possibility of biological cleaning. He had been surprised by the effectiveness of a clay poultice containing urea and glycerol and proposed that microorganisms were at least partially responsible. Kouzeli (1992) has reported favorably on the technique in comparison with pastes based on EDTA or ammonium bicarbonate.

Biological cleaning, in general, has been little researched (Ranalli et al. 1996; Ranalli et al. 2000). Gauri has demonstrated the use of the anaerobic sulfur-reducing bacterium *Desulfovibriode sulfuricans* in removing the black crust on marble (Gauri et al. 1992). He has argued, moreover, that the bacterium was converting calcium sulfate back into the calcium carbonate from which it was originally formed (Atlas, Chowdhury, and Gauri 1988; Gauri and Chowdhury 1988). Konkol has demonstrated that using an enzymatic cleaner derived from the fungus *Trametes versicolor* may reverse biological staining of marble (Konkol et al. 2009). Efforts to remove lichen from concrete through the use of *Thiobacillus* bacteria have been evaluated by De Muynck, De Belie, and Verstraete (2010). Comparison of sulfate-reducing bacteria treatment versus conventional chemical cleaning procedures on a marble element of the Milan Cathedral is reported by Toniolo et al. (2008) and Cappitelli et al. (2007a).

Targeting the Dirt

Gauri's work is interesting because it takes account of the nature of the dirt. It is true that this may be implicit in other cleaning techniques (e.g., the use of complexing agents to increase the solubility of calcium sulfate or the use of hydrofluoric acid to dissolve silica), but it is disappointing that only a few developments in cleaning techniques have flowed out of the extensive studies on black crusts. One example is the work of Vergès-Belmin, Pichot, and Orial (1994) determining the point at which to stop the removal process. Livingston (1992) has studied the solubilities of calcium carbonate and calcium sulfate; Schiavon (1992) has commented on the distribution of calcium sulfate within the pores of stone and on that distribution's implications for water washing; and Skoulikidis and Beloyannis (1984) have attempted to convert calcium sulfate back into calcium carbonate by the use of potassium carbonate, blissfully ignoring the potentially harmful effects of the resulting potassium sulfate. Few other researchers, however, have focused directly on the nature of the dirt deposits in an attempt to develop more effective cleaning techniques. Partially, this has been due to the fact that it is only recently that the complex amalgam of organic fractions contained in patinas and the role microbes play in this ecology have become better known (see the Biodeterioration section in chapter 1 and the Rock Art section in chapter 5).

ACTIVE CONSERVATION: DESALINATION

In situations where soluble salts are a major contributor to decay, it makes sense to try to remove the salts. The word *try* is used deliberately.

The removal of water-soluble salts sounds tantalizingly easy, but it can prove difficult in practice. Salt reduction may be a more appropriate term (Redman 1999; Sawdy, Heritage, and Pel 2008; Pel, Sawdy, and Voroninaa 2010).

Salt reduction is relatively straightforward in the case of small artifacts, which can, for example, be immersed in water or enclosed completely in a poultice, though even here problems can arise through the frailty of the surface or the presence of pigments (Beaubien et al. 1999; Paterakis 1999; Muros and Hirx 2004; Franzen et al. 2008). The real problems start when one attempts to remove salts from the masonry of a building or monument. In an early desalination study, Bowley (1975) demonstrated that it was possible to extract a worthwhile quantity of salt from masonry through the repeated use of clay poultices, although little would be gained in the long run unless one could eliminate the source of further salt. An excellent review (Vergès-Belmin and Siedel 2005) makes it clear that larger-scale masonry desalination needs further study.

Desalination of masonry is usually attempted through the use of poultices, which may consist of a range of materials (e.g., clay, sand, and paper pulp) (Auras 2008). In those instances where calcium sulfate is to be removed, additional materials may be added in order to increase its solubility. Clearly there are overlaps here with cleaning, especially in the removal of black crusts. The additives may include EDTA and its sodium salts, sodium bicarbonate, ammonium bicarbonate, and ammonium carbonate (Maravelaki et al. 1992; De Witte and Dupas 1992; Alessandrini et al. 1993; Leitner 2005; Henry 2006, p. 153). A word of warning may be appropriate: If a limestone is heavily sulfated, the calcium sulfate may be all that is holding it together, and total removal could be disastrous.

An EC project, Assessment of Desalination Mortars and Poultices for Historic Masonry (DESALINATION) 2006–9, has worked to provide a scientific foundation and guidelines for the efficient application of desalination poultices (Bourguignon et al. 2008; Doehne et al. 2008; TU Delft 2009). Principles involve matching the poultice to the pore characteristics of the substrate (kaolin helps with finer stones), preventing rapid drying of the poultice, using less water, and thinner poultices. While counterintuitive, using less water helps remove salts that are near the stone surface and helps avoid pushing the salts deeper into the stone. Some improvements, using finer poultices and both sides of the wall, have also been proposed by other researchers to improve the efficiency of the desalination process (Friese and Protz 1997; Friese, Protz, and Peschl 1997). More recent work has shown that poultice shrinkage and detachment are further important parameters in improving poultice efficiency (Bourgès and Vergès-Belmin 2008a; Bourgès and Vergès-Belmin 2008b; Sawdy, Heritage, and Pel 2008; Heritage et al. 2008).

Desalination efforts often need to be coupled with efforts to reduce the supply of salts, such as the maintenance or installation of a damp-proof course (DPC) at the base of the building foundation (Pinto Guerra 2008; Young and Ellsmore 2008). Installing new DPCs to deal with the accumulation of salts and damp has a mixed record in some church monuments (Henry 2006, p. 277).

Finally, the use of bacteria in desalination may merit further attention. Gauri's use of sulfur-reducing bacteria to eliminate the black crust has already been mentioned, and Gabrielli (1991) gives an anecdotal account of using the reducing atmosphere created by cow dung to convert nitrate salts into elemental nitrogen gas. One wonders, however, if other salts are added at the same time. Removal of salts by microorganisms has also been proposed by Webster and others (Webster, Vicente, and May 2004; Webster and May 2006) as a central part of the EC BIOBRUSH project (BIOremediation for Building Restoration of the Urban Stone Heritage; May et al. 2008). However, these studies found that any effects of the bacteria were masked in many cases by the effect of the material used to apply them and that there were practical problems in supporting the weight of the application material on large areas. One is left with the feeling that additional development is needed before practical biological cleaning can be readily applied. In contrast, biocalcification appears to be at a much higher level of development (see Lime and Biocalcification section below).

ACTIVE CONSERVATION: CONSOLIDATION

Where stone is severely weakened by decay, some form of consolidation may be necessary to restore some strength. Ideally, one might hope to make the stone at least as strong as it was originally (Snethlage 2008; Scherer and Wheeler 2009), so it might resist further decay, but even the strength to resist the battering of the wind or the wing of a bird may be enough to prolong survival.

It all sounds so easy. One just has to find something that will penetrate the decayed stone, binding it together and securing it onto the sound stone beneath (Ginell, Wessel, and Searle 2001). And why stop there? Why not find something that will also protect the stone from further decay? Perhaps it could prevent damage from cycles of salt crystallization. Or perhaps it could make the surface of the stone water-repellent or able to resist hygric swelling. Of course, the treatment will need to be reasonably cheap, easy to apply, and safe to handle. VOC (volatile organic compound) regulations mean that any treatment needs to be formulated to be environmentally friendly. It will need to remain effective for decades at a time, in order to last from one maintenance cycle to the next (often dictated by the cost of scaffolding). The treated stone will need to have much the same moisture expansion, thermal expansion, and elastic modulus as the untreated stone in order to avoid internal stresses and assure compatibility. Ideally, the treatment should work equally well on any type of stone, regardless of the cause of decay. And let's not forget that it must be completely invisible.

Put like this, it sounds absurd to attempt the task. It is like trying to find one pill that will cure all the diseases known to humankind. But this has not hindered the search for an all-singing, all-dancing stone consolidant-cum-preservative. It is a wonder we have made as much progress as we have. An enormous variety of materials have been tried since time immemorial (Barff 1860; Egleston 1886), each with its own advocates (Palmer 2002).

One has to start somewhere, and one of the properties that a consolidant must have is the ability to penetrate the stone. This, in turn, requires a low viscosity and a low contact angle. Next, the consolidant needs to stiffen or set once it is in place in order to strengthen the stone. These requirements can be met in three ways: first, one could think of applying a substance that is liquid at high temperature and stiffens as it cools down—wax for instance. In practice, it is hard to get a low enough viscosity without excessive heat, and wax tends to be sticky and to pick up dirt. The consolidation might become risky in areas having significant exposure to the sun. The second approach is to use a consolidant dissolved in a solvent. One cannot assume, however, that the consolidant necessarily penetrates as far as the solvent, and there is always a danger of the consolidant being drawn back to the surface as the solvent evaporates. Third, one can use a low-viscosity system that undergoes a chemical reaction in situ to give a solid product.

Consolidants are usually applied to the surface of the stone by brush, spray, pipette, or by immersion and are drawn into the stone by capillary action. Domaslowski (1969) experimented with a "pocket system" that was intended to hold the consolidant against the stone, and Mirkowski (1988) has described a system employing bottles to maintain a steady supply of the consolidant at a large number of points. At St. Trophime (Arles, France), consolidant was fed using "intravenous" tubes, allowing a slow drop-by-drop application to the stone surface (Mérindol 1994). Schoonbrood (1993) has developed a low-pressure application technique that maximizes capillary absorption. Vacuum systems may also be used to facilitate penetration into movable objects and ashlars (see, e.g., Hempel 1976; Török 2008). The vacuum system developed by Balfour Beatty Limited (Balvac) for use on monuments (see, e.g., Antonelli 1979) did not find extensive application in practice. Various vacuum systems for sculpture are in use (Pummer 2008), and damage to fragile stone surfaces can be reduced by wrapping them with cotton.

The majority of materials that have been tried as stone consolidants have been organic polymers, but several inorganic materials deserve a particular mention, as their mode of operation is rather different: calcium hydroxide (slaked lime) and barium hydroxide.

Lime and Related Treatments

Nothing could be more natural than putting lime into limestone. The emotive appeal of lime must account for at least some of its popularity. There is, however, a sound rational basis for its use. If a saturated solution of calcium hydroxide is allowed to penetrate into limestone, subsequent evaporation of the solution will lead to the deposition of calcium hydroxide within the stone. This, in turn, will react with carbon dioxide in the air to produce calcium carbonate. This could serve to consolidate the stone, in much the same way as carbonation of calcium hydroxide leads to the hardening of a lime mortar.

This basic chemistry forms the basis of the "lime technique" (Ashurst 1998), which has been used extensively in England and to a lesser extent elsewhere. The technique, in its entirety, can quite transform the appearance of decayed limestone. However, Price, Ross, and White (1988) demonstrated that the lime was deposited largely in the outer couple of millimeters of the stone and that no deep consolidation of the stone could be attributed to the calcium hydroxide. However, it is conceivable that some consolidation could be attributed to the redeposition of calcium sulfate within the stone, a suggestion supported by the apparent effectiveness of distilled water under some circumstances (Clarke and Ashurst 1972). The conclusion of Price, Ross, and White was that the success of the technique was largely attributable to the subsequent use of well-designed mortars, which filled surface fissures and other defects. An alternative suggestion, put forward by R. White (personal communication) and by Anagnostidis et al. (1992), is that the lime is serving to kill bacteria and other organisms and so reduces decay. Krumbein and others (1993) suggest that the observed sterility of marble treated with lime may be due not to biocidal action but to pore closure, which prevents colonization.

Despite the hope that the lime treatment would lead to the deposition of interlocking calcium carbonate crystals, in the manner of lime mortars, the available evidence suggests that it is deposited in an amorphous form that can have little consolidating effect. Tiano and others, however, have proposed a pretreatment based on glycoproteins derived from marine organisms and biomineralization (Tiano, Addadi, and Weiner 1992; Tiano 1995; Tiano 2004). The pretreatment is reported to induce the nucleation of calcite, leading to well-formed crystals that adhere strongly to the underlying stone. More recent work undertaken by Jiménez-Lopez and colleagues (Jiménez-Lopez et al. 2007; Jiménez-Lopez et al. 2008) tested the consolidating effect of soil microbes precipitating calcite in porous limestones.

The lime technique is still in use (Fidler 1995; Brajer and Kalsbeek 1999; Fidler 2002; Woolfitt and Durnan 2002; Oudbashi et al. 2008). However, new nano-lime technology is now available after some years of development (Giorgi, Dei, and Baglioni 2000; Ambrosi et al. 2001; Dei and Salvadori 2006; Adolfs 2007; Ziegenbalg 2008). This technology, which suspends nano-scale calcium hydroxide particles in alcohol, permits deep penetration into stone surfaces. The use of alcohol instead of water limits carbonation by CO_2 before the particles are deposited in the porous stone and facilitates much higher loadings of lime than is possible with aqueous solutions. The method is commercially available and has been used in some specific cases (Howe 2007; Daniele and Taglieri 2010). Future work should include the long-term testing of nano-lime materials, and an EC project on the topic is in progress: STONECORE (Stone Conservation for the Refurbishment of Buildings, http://www.stonecore-europe.eu/; Drdácký, Silzkova, and Ziegenbalg 2009).

Barium Hydroxide

Barium hydroxide is another material with a long pedigree. Chemically, barium compounds and calcium compounds share many of the same characteristics, the one notable difference being the insolubility of barium sulfate as compared with the sparing solubility of calcium sulfate. Barium hydroxide treatments thus have a number of possible objectives, which are not always clearly spelled out. They may serve to convert calcium sulfate to barium sulfate and thereby reduce damage due to the solution and recrystallization of calcium sulfate; they may serve, after carbonation, to deposit a coating of barium carbonate, which will be more resistant than calcium carbonate to acid rain; and they may serve to consolidate the stone through the formation of solid solutions of barium calcium carbonate (Lewin and Baer 1974). The advantages and disadvantages of barium treatments are reviewed by Hansen and others (2003).

A number of techniques have been proposed for introducing the barium hydroxide into the stone. Simple application of barium hydroxide solution appears to be ineffective and led Schaffer (1932, p. 84) to dismiss the process in just seven words: "In practice the method proved a failure." Lewin and Baer (1974), by contrast, described a technique that ensured the slow growth of well-formed barium carbonate crystals within the stone, a technique Lewin was still advocating fifteen years later (Lewin 1988). Schnabel (1992) has cast doubt on the effectiveness of the process when applied by capillarity in situ. More recent work on barium includes "not satisfying" results from Toniolo et al. (2001), good results on Gioia marble (Bracci et al. 2008), and its use as an additive in lime mortars (Karatasios et al. 2007). An EC project evaluating a range of consolidant treatments, including barium hydroxide, found improvements in drilling resistance to a depth of 2 cm in porous limestones (Bracci et al. 2008).

The widest application of barium hydroxide has come in the field of wall paintings, where Matteini (1991) proposed that barium hydroxide treatment should be preceded by the use of ammonium carbonate to dissolve the calcium sulfate (Ambrosi et al. 2000). Barium oxalates and aluminates have also been tested on a range of materials (Matteini and Zannini 2004).

Organic Polymers

From naturally occurring compounds, such as linseed oil and cactus juice, to the synthetic polymers of the twentieth century, somebody somewhere will have tried it as a stone consolidant. Generally speaking, such trials have been on a rather hit-or-miss basis. Materials have been selected more on the grounds of availability than of any predetermined qualities. Provided they will penetrate the stone and then set, they have been worth a try. In a number of cases, the use of incompatible materials on stone has led to a series of difficult and unintended consequences, even with ostensibly removable materials (Nimmrichter and Linke 2008).[1]

While it is easy to sound contemptuous about such an empirical approach, it is hard to see how things could have been any different. Because our knowledge of decay processes is still incomplete, our knowledge of how to combat them is incomplete, as well. Of necessity, we are learning by experience.

The vast majority of researchers believe that stone needs to "breathe." In other words, stone should remain permeable to water vapor, in order to avoid any buildup of moisture and soluble salts (and consequent shear stresses) at the interface between the treated zone and the untreated stone below. Rapid drying of stone surfaces reduces the potential for biological growth and decreases the time of wetness—a parameter associated with damage to stone from air pollution.

Little attention has been given to the distribution of consolidants within stone at the microscopic level, despite numerous photomicrographs taken with the scanning electron microscope. Many authors have been content simply to state that a treatment "lines the pores." Sasse and Honsinger (1991) have described a "supporting corset" model, consisting of an impermeable layer that coats and protects the internal surfaces of the stone, while imparting mechanical strength. Hammecker and others (Hammecker, Esbert Alemany, and Jeannette 1992; Hammecker 1993) describe the use of mercury porosimetry to monitor changes in pore structure due to treatment, but such studies may be hindered by the change in contact angle following treatment.

Little is known about the bonding, if any, that takes place between a consolidant and the substrate, and much is left to chemical intuition. It is widely argued, for example, that alkoxysilanes will form primary chemical bonds to the Si-OH groups on the surface of sandstones, but that they will not be able to form primary bonds to limestones. Lack of bonding need not necessarily mean failure, however, for an unbonded network of consolidant could still provide strength. The stability of polymers used for protective purposes has been evaluated with increasingly sophisticated methods (Gembinski et al. 2000; Chiantore and Lazzari 2001; Favaro et al. 2005), both in the lab and the field, detailing their alteration and loss of efficiency over time.

More needs to be known, not just about stability but also about the molecular structure of the polymer that is deposited within the stone. We speak glibly, for example, about the network polymer that is formed by the hydrolysis and subsequent condensation of tri-alkoxysilanes and tetra-alkoxysilanes. But how many siloxane bonds are formed, on average, by any one silicon atom? What is the structure of the polymer? How is it influenced by the presence of water, of solvents, of salts, or of particular minerals? How does it affect the strength of the polymer? Our present knowledge of consolidants may be likened to folk remedies in medicine. We have gained a lot of experience of what is, and what is not, effective, but we have little understanding of how polymer consolidants work. Once we have a deeper understanding of the properties that are required of a consolidant, we shall be in a better position to synthesize compounds that incorporate those properties.

Alkoxysilanes

The alkoxysilanes and alkyl alkoxysilanes, or "silanes" for short, have undoubtedly been the most widely used stone consolidants over the past twenty years (Snethlage and Wendler 2000; Wheeler and Goins 2005; Price 2006; Wheeler 2008; Scherer and Wheeler 2009). Two compounds, in particular, have been dominant: methyltrimethoxysilane (MTMOS) and tetra-ethoxysilane (TEOS). The silanes are hydrolyzed by water to form silanols, which then polymerize in a condensation reaction to give a silicone polymer. The water may come from the atmosphere or from the stone itself, or it may be added as a deliberate ingredient. In the latter case, a solvent may be required in order to make the mixture miscible. A catalyst may also be added, usually in the form of an organo tin or lead compound. The condensation reaction, and often the hydrolysis reaction also, takes place after the treatment has been absorbed by the stone, and the resulting polymer imparts the required strength to the stone.

The popularity of MTMOS and TEOS is no doubt due in part to their commercial availability, and a number of proprietary products are available that are based on these two compounds. A number of other silanes have also been tried, usually involving substitution of the methyl group for larger alkyl or aryl groups.

A thoughtful review by Wheeler (2008) of the use of alkoxysilanes for stone consolidation deals with three important issues: the use of alkoxysilanes on clay-rich stone, alkoxysilanes used on limestone versus quartz sandstones, and the use of alkoxysilanes on marble. Results for clays are mixed: two important studies found that ethyl silicate treatment of clay-rich stone initially resulted in a strength increase, but that this improvement was lost after three to ten wet/dry cycles (Félix 1996; Scherer and Jiménez-González 2008). This suggests that for clay-rich stone, the focus should be on reducing clay swelling, not on increasing strength (see the Differential Stress section in chapter 1 for more on anti-swelling treatments).

The difficulty of bonding a silicate material to calcite has long been considered an important problem, resulting in some new research on coupling agents and alternative consolidants (Wheeler, Mendez-Vivar, and Fleming 2003; Correia and Matero 2008; Ferreira Pinto et al. 2008; Ferreira Pinto and Delgado Rodrigues 2008). Wheeler (2008) points out that while the percent strength increase for limestone after ethyl silicate treatment is not as great as for sandstone, comparing the absolute level of the modulus of rupture (generally higher for limestone) provides a more realistic perspective and helps explain the widespread use of this material on limestone. The use of alkoxysilanes on marble is explained as filling narrow voids between calcite grains, which can help lock in particles experiencing granular disintegration (Ruedrich, Weiss, and Siegesmund 2002).

Recent work on nano particle–modified silanes show they reduce the cracking seen in conventional treatments and result in improved consolidation (Escalante, Valenza, and Scherer 2000; Miliani, Velo-Simpson, and Scherer 2007; Kim et al. 2008). Elastified silanes have also been developed to help create a less brittle film (Boos et al. 1996; Kim et al. 2008; Maravelaki-Kalaitzaki et al. 2008). A commercial elastified version is available (E. Wendler; Remmers KSE 500 E). Surfactants have also been tested and result in a less brittle silane treatment—a hybrid nanocomposite (Mosquera and de los Santos 2008; Simionescu et al. 2009).

Important research on application procedures has shown that the timing and number of applications can result in important differences in the pore-blocking effect and general hardness of TEOS (De Clercq, De Zanche, and Biscontin 2007). The development of microporosity during the curing of Funcosil stone strengthener was noted by Barajas and others (2009).

Although the literature contains many papers describing the use of silanes on stone, there are few that attempt to come to grips with the underlying chemistry or the associated sol-gel technology. Some exceptions are studies by Wheeler (Wheeler, Mendez-Vivar, and Fleming 2003; Wheeler and Goins 2005; Scherer and Wheeler 2009), Scherer (Scherer, Flatt, and Wheeler 2001; Miliani, Velo-Simpson, and Scherer 2007), and Snethlage (Snethlage 2002; Meinhardt-Degen and Snethlage 2007; Snethlage 2008). Other recent work includes efforts to evaluate and control the relationship of pore evolution and solvent (Salazar-Hernández et al. 2009). Research continues on extending sol-gel treatments beyond stone to other diverse heritage materials, including bronze, pyrite, and unstable historic glass (Bescher and Mackenzie 2003; Khummalai and Boonamnuayvitaya 2005; Dal Bianco and Bertoncello 2008). Kumar (Kumar and Price 1994) has reported on the influence that soluble salts may have on the hydrolysis and condensation of MTMOS. Sodium sulfate, for example, markedly decreased the rate of both hydrolysis and condensation, whereas sodium chloride increased the rate of condensation. Silica-sol treatments at Petra were found to perform poorly in the presence of salts, resulting in the need to poultice areas to be treated prior to application (Simon, Shaer, and Kaiser 2006). Consolidation of stone does not encapsulate salts that may be present, and research shows that salts can be removed by poulticing after treatment if salt concentrations are low to moderate. However, some of the consolidation effect is lost after wetting the samples, depending on the salt tested (Costa and Delgado Rodrigues 2008a).

Epoxies

Epoxy resins have had some bad press as far as consolidation is concerned. Many conservators see them as viscous, brittle, yellowing materials that may make admirable adhesives in some circumstances, but which are certainly not to be considered as consolidants.

It is true that there have been some notable failures, but it would be foolish to dismiss epoxy resins entirely on these grounds. Selwitz, in three reviews (1991, 1992a, 1992b), summarized the use of epoxies as consolidants, charting the successes and failures. He highlights the pioneering work of Domaslowski (Domaslowski and Strzelczyk 1986; Domaslowski and Sobkowiak 1991) and Gauri (1974; Gauri and Appa Rao 1978) and emphasizes the two different paths they have adopted in order to treat relatively small objects and large facades, respectively. The choice of solvent, the means of application, and postapplication procedures are vitally important to a successful outcome (Pinto and Delgado Rodrigues 2008b).

Cycloaliphatic epoxy resins (Eurostac EP2101) have been successfully used in some important field consolidation cases in Italy, such as the deep consolidation under vacuum of large, fissured granite columns (Cavalletti et al. 1985). More recent work has focused on application methods that minimize color change with aging (Ginell and Coffman 1998), the use of waterborne epoxy emulsions (Kozub 2004; Luan Xiaoxia et al. 2008), and complex hybrids, such as epoxy-silica materials (Cardiano et al. 2003; Cardiano et al. 2005). Further work is needed to evaluate these newer materials.

Acrylics

Although in situ polymerization of methyl-methacrylate (and other acrylic monomers) has its advocates, the high rigidity and glass transition temperature of polymethyl-methacrylate are generally considered to make it unsuitable as a stone consolidant. Far more attention has been given to the use of acrylic resins dissolved in solvents, and the ubiquitous Paraloid B72 (Acryloid B72) inevitably makes its appearance.

Many conservators have experimented with B72 dissolved in an alkoxysilane such as MTMOS, the reasoning being that the B72 brings adhesive properties that the alkoxysilane lacks. The idea seems to have been that B72 is capable of securing pigment or loose flakes, for example, while the alkoxysilane provides deep consolidation. This treatment was used by Nonfarmale and Rossi-Manaresi in San Petronio Cathedral in Bologna, from where the term "Bologna cocktail" was coined (Gnudi, Rossi-Manaresi, and Nonfarmale 1979). In San Petronio, the limestone is very compact and virtually nonporous, and the decay progresses mainly with the formation and detachment of scales and other fragments. The cocktail was used in this case for gluing the scales, and because it was properly done by a very experienced conservator, the result was satisfactory and those surfaces are apparently still in good condition (Laurenzi Tabasso 1995). The problem arises when the Bologna cocktail is transposed to very porous limestones with pore-shaped voids (J. Delgado Rodrigues, personal communication). Under these circumstances, B72 has a very low impregnation capacity, forming indurated crusts and leading to severe detachments some time after application. In such cases it may constitute a disaster. In short, Paraloid B72 is an excellent adhesive, but it is not necessarily a good consolidant outdoors. The Bologna cocktail is a useful example of the need to match the treatment to the problem and the need for critical thinking when navigating the conservation literature. More recent research on the aging of Bologna cocktail mixtures (Paraloid B72 and Dri Film 104) has been undertaken by Favaro and others (2006; 2007).

Wheeler and co-workers (Wheeler et al. 1991; Wheeler, Wolkow, and Gafney 1992) have shown that the resulting composite gel is weaker than the polymers derived either from neat MTMOS or from a solution of B72 in a nonreactive solvent. Research has continued on acrylic/siloxane composites (Zielecka, Bujnowska, and Bajdor 2007; Sadat-Shojai and Ershad-Langroudi 2009) with some promising results. Other work on B72 has focused on characterizing its long-term stability and field performance (Roby 1996; Bracci and Melo 2003).

Other Materials

Innovative approaches to consolidation have come from several researchers, such as the use of calcium alkoxides (Favaro et al. 2008), the transformation of gypsum or calcite into calcium phosphate based on historic patinas (Martín-Gil et al. 2005; Xiangmin Zhang and Spiers 2005; Vazquez-Calvo, Alvarez de Buergo, and Fort 2007; Snethlage et al. 2008), and frontal (in situ) polymerization (Proietti et al. 2006; Mariani, Capelletti, and Brunetti 2008).

Research on the use of tartrates has led to a patented product that creates a conversion coating on calcite that can also act as a coupling agent for ethyl silicate–based treatments (Slavid and Weiss 2001). Known commercially as HCT (Prosoco, Inc.), the product has been on the market for some time (Correia 2005; Correia and Matero 2008; Pinto and Delgado Rodrigues 2008a), and results from longer-term trials are expected in due course.

Previous accounts of isocyanates, polyurethanes, and polyureas may be found in Hansen and Agnew (1990); Coffman, Agnew, and Selwitz (1991); Zádor (1992); Littmann et al. (1993); Auras (1993); and Riecken and Sasse (1997). The use of cyclododecane, largely as a temporary, reversible consolidant that sublimes over time, has been explored over the past decade as a useful new component of the conservator's toolbox (Stein et al. 2000; Maish and Risser 2002; Muros and Hirx 2004; Anselmi, Doherty, and Presciutti 2008). Some health and safety issues regarding cyclododecane remain to be resolved (Rowe and Rozeik 2008). Advanced research in self-cleaning surfaces, such as titanium-coated glass, has led to interest in biomimetic surfaces that may have potential application for developing compatible coatings for the conservation of stone (Solga et al. 2007; Qiang Liu et al. 2006; Kun Hong and Yuzhong Zhan 2008).

Emulsions

Organic consolidants frequently rely on the loss of volatile reaction products or solvents during the curing process. This can make application impracticable in hot climates, and it can pose a hazard both to the conservator and to the wider environment. Attention has been given to the development of aqueous emulsions for use as consolidants and as surface coatings (see the following section). Snethlage and Wendler (1991) discuss the possible use of an aminoalkyl silane to stabilize a silane emulsion, and Piacenti, Camaiti, Brocchi, and others (1993) report on the development of emulsions based on a hexafluoropropene-vinylidene fluoride elastomer. More recent work illustrates the diverse application of emulsions containing acrylic, fluorinated acrylic, methacrylate/alkoxysilane, or epoxy resin as conservation treatments (Castelvetro et al. 2004; Luan Xiaoxia et al. 2008; Theoulakis et al. 2008). Further work in this area seems probable.

SURFACE COATINGS

Surface coatings is a bit of a catchall category that includes a range of materials applied to stone—protective water repellents, emulsions, antigraffiti coatings, salt inhibitors, protective oxalate layers, sacrificial lime coatings, colloidal silica, biocides, and bioremediation treatments. A substantial research effort in the 1970s and 1980s was aimed at finding a single treatment that would both consolidate and protect stone. However, the naïveté of this approach has become increasingly apparent, and many conservators now accept the need for two treatments: one to consolidate and one to protect. The soundness of the latter approach has been borne out by Félix and Furlan (1994) and Alonso and others (1994), who reported damage to certain stones following treatment with tetra-ethoxysilane (TEOS) unless the stones were also given a water-repellent coating.

Protective treatments need to be maintained, and this means retreatability needs to be taken into consideration when designing a treatment system. Surface coatings can be renewed at regular intervals, but the initial consolidation will, it is hoped, last much longer.

Some researchers have suggested doing away with the consolidant and relying solely on the water repellent (Sramek 1993). However, the long and disappointing history of water-repellent coatings on the more porous limestones and sandstones should not be dismissed too readily (Honeyborne et al. 1990).

Water Repellents

The property that has been most sought in surface coatings is water repellency. The logic behind the approach is simple: Since water is involved in most forms of stone decay, a treatment that prevents the ingress of water should help to reduce decay. Reviews of the use of waterproofing agents on stone can be found in Charola (1995), Bromblet and Martinet (2002), as well as Vallet and others (2000). The influence of the substrate and the temperature of application for water repellents have been investigated by De Clercq and De Witte (2001). A series of conferences on water repellents have been held, with the most recent in Brussels, Belgium in 2008 (International Conference on Water Repellent Treatment of Building Materials: Hydrophobe V 2008), (De Clercq and Charola 2008).

Water repellency has been provided largely by alkoxysilanes, silicones, and fluoropolymers. The development of the fluoropolymers provides an interesting, and regrettably rare, instance of "tailor-made" products. The polymers are close relatives of polytetrafluoroethene (PTFE, or Teflon), renowned for its nonstick properties. The early fluoropolymer coatings worked well, except for a rather poor ability to stick to the stone! Subsequent development has entailed the synthesis of compounds containing functional groups that can adhere to the stone surface, thereby providing more persistent protection (Piacenti, Camaiti, Manganelli del Fa, et al. 1993). It has been argued that such water repellents should help to prevent resoiling, although this claim has not been adequately substantiated. The rapid loss of water-repellent properties after accelerated (artificial) and field (natural) weathering has been noted by several researchers and deserves further study.

Another example of "tailor-making" is provided by Fassina and co-workers (Aglietto et al. 1993; Fassina et al. 1994), who have synthesized a range of fluorinated acrylic polymers. The intention, which was partially achieved, was to improve water repellency and resistance to photooxidation, by comparison with nonfluorinated analogues such as Paraloid B72. In a different approach, research on the use of polyurethanes on stone, known as the "Aachen concept," has been reviewed by Snethlage and Wendler (2002).

More recent work on several types of water repellent (an acrylic dispersion, an oligomeric alkylpolysiloxane, a solution of silicone resin and an alkylalkoxysiloxane in aqueous microemulsion) applied to seven types of limestone found that: "Due to the diverse petro-physical nature and properties of each stone, the results indicate that no universally compatible protective treatment exists" (Boutin 2001, 233). Accelerated or artificial tests of hydrophobic coatings as a method for reducing the effects of air pollutants on porous, calcareous stone have had mixed results, with the protective effect decreasing rapidly with time in bulk samples (Camaiti et al. 2007), while X-ray photoelectron spectroscopy (XPS) microanalysis showed adequate performance after aging 240 hours (Torrisi 2008).

In addition to surface layers on stone, the water-repellent properties of silanes have also been used to create chemical damp-proof courses (DPC) along the base of foundations of buildings that lack this common feature of modern masonry buildings (Pinto Guerra 2008; Young and Ellsmore 2008). While an ancient idea (see Vitruvius 7.4), DPCs began to be standardized in new construction only starting in the mid-nineteenth century (Schmidt 1999). Chemical DPC application methods have included gravity feed and pressure injection of silanes into regularly spaced holes drilled into the foundation. Current methods include a cream containing silane injected into holes drilled along a mortar joint. The silane apparently diffuses out of the cream and some distance into the mortar to form a chemical DPC. The long-term performance of various DPC measures suggests that some may experience a rapid loss of effectiveness over time (Alfano et al. 2006; Lopez-Arce et al. 2009; Henry 2006, p. 277).

Recent work shows that sodium chloride preferentially crystallizes on hydrophobic surfaces (Shahidzadeh et al. 2008), suggesting that water repellents are not compatible where salts may accumulate (Lubelli et al. 2007). An EC-funded project, SCOST (Salt Compatibility of Surface Treatments), addresses this issue in detail (De Witte 2001; Miquel et al. 2001).

Anti-Graffiti Coatings

The problem of graffiti has spread across diverse urban environments over the past fifteen years and is affecting not just modern buildings but historic monuments as well. A new EC project on the topic, comparing five graffiti protectives in six countries (Gardei et al. 2008), has found that four commercial anti-graffiti agents strongly reduce water and vapor transport and thus are not compatible with most historic building materials. However, a new product developed specifically for historic materials was found to have acceptable performance and is undergoing field testing. Recent work on an anti-graffiti coating containing perfluoropolyether and epoxysilanes in aqueous microemulsion with an epoxide curing agent found good resistance to repeated cleaning cycles (Licchelli and Marzolla 2008). Earlier work by Mertz, Grunenwald, and Ternay (2003) found that some reduction in water vapor permeability was necessary to get efficient protection and that the preventive anti-graffiti treatments do not perform the same on substrates with high and low capillary absorption coefficients.

Emulsions

Complex emulsions as stone protectives have been studied by a number of researchers. The emulsions have included acrylics (Kumar and Ginell 1995; Theoulakis et al. 2008; Karatasios et al. 2009), silicones (Snethlage and Wendler 1991; Ren and Kagi 1995; Mao and Kagi 1995; Van Hees and Koek 1995; Ciabach 1996; Boutin 2001), silanes (Biscontin et al. 1993; Licchelli and Marzolla 2008; Wittmann et al. 2008), and fluorinated polyurethanes (Guidetti, Chiavarini, and Parrini 1992; Croveri and Chiavarini 2000). Performance varies from stone to stone but is generally promising.

Crystal Growth Inhibitors

Another possibility is to treat the stone surface with compounds that inhibit the growth of salt crystals, as was mentioned briefly in the section on salts. Relevant technology already exists in such diverse fields as anti-caking agents for road salt and in oil extraction, where phosphonates are used to prevent the precipitation of barium sulfate and calcium sulfate (Black et al. 1991). Applications in the field of conservation have been proposed from time to time (e.g., Puehringer and Engström 1985), and recently this area has received some further research (Selwitz and Doehne 2002), including an EC project, SALTCONTROL, on the topic (Rodríguez-Navarro, Hernandez, and Sebastian 2006; Cassar et al. 2008). Inhibitors used to treat stone surfaces, such as phosphonates and carboxylates were found to be a mixed blessing. In some instances they decrease damage by letting salts reach the surface as less harmful efflorescence. However, in other situations they enhance solution supersaturation ratios and absorb to surfaces, resulting in increased rates of damage.

Oxalate Formation

Building on the protective properties of *scialbatura* (see the Biodeterioration section in chapter 1), Matteini, Moles, and Giovannoni (1994) tested the use of ammonium oxalate to produce a shallow film of calcium oxalate on calcareous surfaces such as wall paintings. Both calcium carbonate and calcium sulfate react with a poultice containing a solution of ammonium oxalate to produce a cohesive, hydrophilic film that reduces rates of acid attack (Hansen et al. 2003; Doherty et al. 2007; Sikka et al. 2008). The method has been used to help protect objects and surfaces that cannot be removed to a more protective environment (Ambrosi et al. 2000; Mairani, Matteini, and Rizzi 2000).

Lime and Biocalcification

The final stage of the lime treatment consists of the application of a very thin coating of lime and fine aggregates rubbed firmly into the surface of the stone (see above under the Lime and Related Treatments section). The coating is intended to protect the stone, and it is reapplied as necessary. An alternative approach, which started first in France, utilizes microbes to produce a sacrificial surface layer of calcite (Orial et al. 1996; Le Métayer-Levrel et al. 1999; Castanier et al. 2000; Orial, Vieweger, and Loubiere 2003). Results from an EC project on bioremediation (BIOBRUSH) have been presented by Webster and others (Webster, Vicente, and May 2004; Webster and May 2006; May et al. 2008). Biocalcification in the context of conservation treatments is reviewed by Tiano (2008), and promising test data is presented by Zamarreño, Inkpen, and May (2009).

Colloidal Silica

Kozlowski, Tokarz, and Persson (1992) have adopted a rather different approach for forming a protective coating on calcareous stones. They have used sols of colloidal silica that deposit silica particles within the outer pores of the stone. The resulting surface is hydrophilic, but the passage of water through the pores is impeded by the presence of the particles. The material has been used at several sites to help protect vulnerable calcareous materials from acidic pollution (Stepien, Kozlowski, and Tokarz 1993; Mangio, Simpson, and Tokarz 1996). The method has been further developed by the conservator Egon Kaiser for use as a void filling and repair mortar at Petra and other sites (Kühlenthal, Kaiser, and Fischer 2000; Simon, Shaer, and Kaiser 2006).

Biocides

There is a long history of research into surface treatments that will kill biological growths and, if possible, inhibit regrowth. Such treatments must meet a large number of criteria, and this can prove difficult in the outdoor environment, where there is a continual supply of moisture to promote regrowth. They must not only kill the growth in the first place but also be resistant to new strains. They must not have any harmful effect on the stone itself, nor must they change its appearance. They must not be washed out by rainfall before taking effect or destroyed by ultraviolet light, and they must be safe both to the person applying them and to the wider environment. The last requirement has been applied evermore stringently over the past few years, with the result that a number of proven biocides have been banned by law. It follows that there is still a need for research in this area. The related area of biological stain removal has seen some development and success in removing some stubborn materials (Delgado Rodrigues and Valero 2003; Konkol et al. 2009).

Most of the existing research on biocides has been concerned with algae, lichens, and higher plants like weeds, mosses, and ivy. Some of the research has been on cultures in the laboratory, while most of it has been based on site trials. Examples of such research are provided by Agarossi, Ferrari, and Monte (1990); Monte et al. (2000); and Anagnostidis and others (1992). The last also emphasize the need for regular observation and retreatment, and they suggest early warning systems to indicate the moment for retreatment. A promising new approach to biocontrol using anti-biofouling agents is presented by Cuzman, Tiano, and Ventura (2008). An interesting example of a complex fungal treatment is outlined by Orial and Brunet (2004), while a recently proposed treatment for lichens is removal by a low-pressure abrasive technique using dry ice (Rosato 2008). Laser treatment for lichens has been investigated by DeCruz and others (2009). A book reviewing the topic of biocides for natural and artificial stone is in preparation (Daniela Pinna, personal communication).

Caneva, Nugari, and Salvadori (1991; 2008) provide a valuable account of the many available biocides, which are normally applied to the surface of the stone by brush or spray. Portable objects may also be treated by fumigation: Elmer and others (1993), for example, report the use of ethylene oxide. Bassier (1989) reports the use of ultraviolet radiation to sterilize mineral surfaces. Caneva, Nugari, and Salvadori (1991, p. 119; 2008) mention the possibility of preventive conservation by the deliberate introduction of suitable vegetation in the vicinity. Some water-repellent treatments act to prevent biological growth by limiting available water. Low tech is still a useful approach, as shown by work using hot water vapor to kill lichens and algae (Orial and Bousta 2005).

Sorlini, Falappi, and Sardi (1991) report the inhibition of fungal growth by a methylphenyl silicone resin, but other workers (Petushkova and Grishkova 1990; Santoro and Koestler 1991; Krumbein et al. 1993) have reported the opposite effect: the biodegradation of silicones.

Relatively little research has been conducted on antibacterial treatments for stone. This is surprising, perhaps, in view of the extensive work on the role of bacteria in decay, but it may reflect the difficulty of finding antibacterial treatments with sufficient persistence (Gorbushina et al. 2003). Nonetheless, Orial and Brunet (1992) present a satisfying account of the use of streptomycin and kanamycin to substantially reduce bacteria in stonework at Elne Cathedral for a period of more than seven years, with a resulting cessation of decay.

Biological Attack on Treatments

In some cases polymeric treatments of stone become food for microbes, leading to the production of organic acids and other biological activity related to the consumption of surface treatments (Cappitelli et al. 2007b; Cappitelli and Sorlini 2008). However, this biological affinity for certain otherwise insoluble, cross-linked organic material has also been used as a bioremediation treatment to remove the hardened glue from the surface of a fresco fragment in storage for twenty years (Antonioli et al. 2005).

Note

1 This was also discussed in a paper presented by Simon Warrack at the Stone Consolidation in Cultural Heritage: Research and Practice Symposium, held in Lisbon, May 6–7, 2008.

Chapter 3

Do They Work? Assessing the Effectiveness of Treatments

One might suppose that the most practical approach to stopping or reducing stone decay would be simply to apply a treatment and see if it works. But how can we tell if it is working? What do we really mean by "working"? How long does a treatment need to be left in place? Can things be speeded up a bit? Will it keep on working indefinitely? Will it work on other stones in other environments? What about other treatments that come along while a lengthy evaluation of one is being carried out?

Price (1982) reviewed strategic approaches to the evaluation of treatments, an issue that lies at the crux of the conflict between "doing something" and not causing harm. It is a subject that is of vital importance. We need answers straightaway in order to devise responsible programs for the conservation of monuments that are decaying before our eyes. But if we act too quickly and apply the wrong treatment, we may make matters even worse.

Many researchers have devised their own procedures for evaluating treatments, using a range of tests to build up an overall picture (Sasse and Snethlage 1996; Van Hees 1998; Moropoulou et al. 2000; Haake, Simon, and Favaro 2004; Laurenzi Tabasso and Simon 2006; Bracci et al. 2008; Costa and Delgado Rodrigues 2008b). This is both inevitable and understandable, since individual researchers are constrained by the range of techniques that are available to them. Having a range of techniques also has the advantage that the procedure can be tailored to suit a particular stone and environment (Galán and Carretero 1994). It is unrealistic to think that any single procedure could fit all circumstances. However, it can be very difficult to compare the findings of one researcher with those of another, and there is a need for standardized procedures. This was the underlying objective of the RILEM (Réunion Internationale des Laboratoires et Experts des Matériaux, systèmes de construction et ouvrages) Commission 25-PEM (Protection et érosion des monuments) Working Group, albeit not fully achieved (RILEM 1980; Price 1982). The definition of individual test methods has had more success (see the following), although the "not invented here" syndrome frequently hinders their widespread adoption. Some useful advances in standardized evaluation procedures have been made by CEN (Comité Européen de Normalisation) Technical Committee 346 (Fassina 2008).

It is convenient to divide evaluation procedures into two categories: those that characterize the stone shortly after treatment has taken place, and those that are concerned primarily with monitoring long-term performance. Questions that should always be asked before proceeding are: "What criteria apply, and is enough information available to sustain a recommendation to use this or that stone treatment?"

CHARACTERIZING THE TREATED STONE

There are some properties that are helpful in building up an overall picture of the treated stone, even though they may not give any direct indication of the treatment's effectiveness. These include the porosity and pore size distribution of the treated stone, its appearance, and the depth of the treatment's penetration. The majority of tests, however, are concerned with measuring properties that are known to change as the stone decays or with assessing the extent to which the treatment has met certain clear objectives.

Properties That Change with Decay

We have already looked at a number of tests that are intended to measure the extent of stone decay. These tests can obviously be applied to stone that has been treated with a consolidant or surface coating in order to determine whether there has been an improvement in performance. Such tests might include water uptake, Scotch tape tests, surface hardness, drilling resistance profile, and ultrasonic pulse velocity (Giorgi, Dei, and Baglioni 2000; Vergès-Belmin and Laboure 2007). The paper by Villegas, Vale, and Bello (1994) illustrates the difficulties of interpretation that may arise. If it is difficult to characterize decayed stone, then characterizing treated stone is doubly difficult. And if one treatment is difficult enough to characterize, work on stone treated with both consolidant and waterproofing agents shows that together they have a larger effect on pore structure than any single treatment (Iñigo et al. 2001). Then consider the important issue of retreatment of treated stone.

Meeting Objectives

Some tests are designed to assess the extent to which the treatment is meeting certain objectives. If the treatment is intended to impart water repellency, for example, then measurements of contact angle and water absorption are appropriate. If it is intended to provide protection against acid rain, then measurements of weight loss or of salt formation may be necessary. The logical outcome of this approach is the definition of a set of performance criteria against which a treatment may be judged; Sleater (1977) provides a good example. However, Sleater was unable to find any treatment that met all his criteria, and this may account in part for the lack of attention subsequently given to establishing overall criteria.

Caution must be exercised when using tests intended primarily for untreated stone. For example, a crystallization test that relies on the absorption of a sodium sulfate solution is frequently used to determine a stone's resistance to salt weathering (Doehne 2002). The test should not be applied unthinkingly, however, to a stone that has been coated with a water repellent. Excellent performance in the test (e.g., Villegas and Vale 1992, p. 1259) would not necessarily indicate increased resistance to salt growth; it could simply indicate that the water repellent had prevented the ingress of salt in the first place.

Standard Test Methods

Standard test methods are essential if one is to compare the results of different laboratories in any meaningful way. Even for a seemingly straightforward property such as water repellency, differing procedures yield differing results (Henriques 1992). Subsequent interlaboratory testing of some hygric European Norm (EN) standards has found that while many work well, vapor transmission tests were found to have large variations when done according to EN norms (Roels et al. 2004).

The standardization of test methods has been the objective of both national and international committees. Notable among them are the recommendations of the RILEM 25-PEM and 59-TPM (Traitement des monuments en pierre) Working Groups (RILEM 1978; Pien 1991) and the standards published by the Italian Commissione NORMAL (Alessandrini and Pasetti 2004). It is regrettable that details of these test methods have not been more readily available and more widely translated. However, European standards for stone conservation are currently being integrated under the EN norms, with CEN Technical Committee 346 being led by Fassina (2008). The work incorporates the Italian NORMAL, German DIN, RILEM, and other national standards groups (Koestler and Salvadori 1996; Alessandrini and Laurenzi Tabasso 1999; Fontaine, Thomson, and Suter 1999). Other work in building materials standards can be found in various ASTM (American Society for Testing and Materials) and RILEM committees (http://www.astm.org; www.rilem.net).

The trend of research on standards has been to find and define quantitative parameters to characterize materials and help guide treatments as a way to ensure compatibility between interventions and existing materials (Sasse and Snethlage 1997; Bromblet et al. 2002; Laurenzi Tabasso and Simon 2006; Laurenzi Tabasso 2008). However, the extensive list of parameters that researchers suggest should be measured to evaluate treatments has been shown to be unrealistic in the field (Moraes Rodrigues and Emery 2008). As a practical matter, most treatment evaluations have focused on changes in water uptake, color, ultrasonic measurements, or drilling resistance profiles (Ferreira Pinto and Delgado Rodrigues 2008).

LONG-TERM PERFORMANCE

It is one thing to find a treatment that performs well in the short run; it is another thing altogether to be sure that it will keep on performing year after year when exposed to the weather. When a water repellent progressively loses it hydrophobic properties, we may say that its effectiveness is decreasing. However, when the application of a stone consolidant leads (with time) to differential stress and the eventual detachment of the indurated scales, we may say that it is showing a "delayed harmfulness." Both are examples of long-term performance, yet they represent two distinct phenomena.

Natural exposure trials provide the only true test. They may be carried out in situ on monuments or on small blocks of stone that can be brought into the laboratory at intervals for evaluation. Either way, the trials can provide information only on a limited number of stones, treatments, and environments, and it may be many years before reliable information is obtained. A new range of treatments will inevitably have emerged in the meantime. Additionally, one is still confronted with the difficulty of evaluating the effectiveness of the treatment. The surface may look sound on the outside, but what is going on underneath? In situ monitoring relies heavily on the techniques described in chapter 1.

Nevertheless, there are a number of important questions that have been asked and answered by long-term natural exposure trials. For example, Moreau and others (2008) asked the question "Do water-repellent treatments reduce soiling in protected parts of monuments, and do they allow for easier cleaning?" They found after a ten-year study that silicone water-repellent treatments did not decrease the limestone soiling rate, while a fluorinated acrylic resin decreased it significantly. This result is encouraging and suggests that new-generation fluorinated acrylic resins could be used to protect stone against soiling. Not all fluorinated acrylic resins are suitable for every type of stone, however, since some are film forming and may peel away. Sulfation rates were not decreased by water-repellent treatments. After measuring soiling and sulfation, the test slabs were cleaned by micro sandblasting and laser to determine if the coatings had changed the cleaning efficiency. The results show that treated samples typically were not easier to clean by micro sandblasting but instead became lighter in color than untreated samples. After laser-cleaning treated and untreated samples, the typical yellowing observed after laser cleaning was less noticeable on samples treated with silicon-based water repellents. The study also showed that the darker the samples were after exposure, the yellower they were once laser-cleaned (Moreau 2008).

Another interesting example of long-term analysis of treatment performance is research by Favaro and others (2005; 2006; 2007), where they analyzed the effectiveness (1979–2005) of consolidants and water repellents on marble in the field in Venice and in the laboratory. One of the findings was that Paraloid B72 undergoes irreversible modifications over time and becomes impossible to remove completely. Thus, while a treatment may be technically removable at the time of application and in compliance with the dictum of reversibility, over time this may not be the case.

Accelerated weathering chambers are extensively used to simulate decay (see, for example, Sasse and Riecken 1993), but they also introduce another layer of uncertainty (Warke and Smith 1998). Do they accurately reflect long-term behavior? By what factor do they increase the rate of weathering and decay? When consolidated stones are subject to artificial aging, it is common to call these trials "durability" tests. However, what these trials essentially assess is not the durability of the consolidation properties but the delayed harmfulness introduced when a consolidated layer is present. A stone consolidant may keep its strengthening properties intact and show poor performance—delayed harmfulness—in an aging test.

It is noteworthy that most of the literature on long-term performance is concerned with the behavior of the treated stone per se. There have been surprisingly few in-depth studies of the breakdown of the treatment itself. Even in the field of alkoxysilane consolidants, few authors have made systematic studies of the long-term weathering of the resulting silicone polymer (Favaro et al. 2006). Some authors, however, have highlighted the fact that treatments may serve as an energy source for microorganisms (Koestler and Santoro 1988; Petushkova and Grishkova 1990; Krumbein et al. 1993; Cappitelli et al. 2002; Cappitelli and Sorlini 2008). This important aspect had been largely overlooked hitherto.

Documentation of Field Trials

There is a regrettable tendency for researchers to set up field trials, to monitor them for a few years, and then to forget about them as further treatments become available. There is a need for systematic, long-term monitoring of trials. This is often hindered, however, by woefully inadequate or missing records.

The availability of sophisticated databases offers the possibility of creating good, centralized records, and this possibility has been seized by a number of workers. Fitz (1991; 1996) described the MONUFAKT database adopted by the German federal environmental agency, and Rosvall and Lagerqvist (1993) developed the EUROCARE database. More recent work on databases has focused on regional issues (Klamma et al. 2006; Hyslop et al. 2009) and specific projects (Inkpen et al. 2004; May et al. 2004; Cassar 2004). The difficulty lies in persuading researchers to put reliable, comprehensive information into the system and in persuading others to use it in future years. This is on top of the difficulty of curating digital records over long time periods, during which hardware, software, and even institutions may rapidly become unsustainable. There is now a movement to encourage researchers to put machine-readable data directly onto the Internet (Grossenbacher 2009; Rosling 2009). Thus, while the technology is available and the potential benefits of data sharing are evident, this approach to databases has seen little implementation and sustained effort, aside from project- or region-specific efforts (GCI 2009).

Chapter 4

Putting It into Practice: Conservation Policy

Conservation is not immune to the vagaries of fashion—fashion that varies with both time and place. In England, for example, it was fashionable one hundred years ago to replace decayed sculpture with "copies"—contemporary interpretations of what the originals might once have looked like. By contrast, the current normal philosophy is to "conserve as found"—to keep original material and prevent further deterioration as far as is practicable. A further approach is common in the Far East, where the emphasis is more on preserving the function of a monument than on preserving the materials from which it is constructed.

Numerous attempts have been made to codify conservation principles and to introduce international uniformity. Notable among them are the 1964 International Charter for the Conservation and Restoration of Monuments and Sites (the Venice Charter), the Burra Charter (Australia, ICOMOS 1999), and *Principles for the Conservation of Heritage Sites in China* (China, ICOMOS 2000). It is beyond the scope of this publication to discuss conservation principles in detail, but it is relevant to note that many parties play a role in shaping conservation policy: the architect, the art historian, the scientist, the archaeologist, the conservator, the owner, and ultimately, the general public. The scientist may be convinced of the validity and importance of his or her results, but there are others to be convinced before the results can impact on conservation policy. One example of an interesting policy discussion of how research in heritage science is organized and carried out on a national level took place in 2005–6 in the UK (House of Lords, Science and Technology Committee 2006; House of Lords, Science and Technology Committee 2007) (see also: http://www.heritagescience.ac.uk).

This chapter focuses on just three aspects of conservation policy: the responsible use of surface coatings, adhesives, and consolidants; the problems posed by multiple treatments; and recording. These three issues have been chosen because they have a common thread, which is the fact that no treatment can be expected to last forever. However much we may be lured into thinking that a treatment will last indefinitely (or, perhaps, until we are no longer accountable for it?), we must accept that all treatments have a finite life. This has direct implications for conservation policy in the three areas indicated.

RESPONSIBLE USE OF SURFACE COATINGS AND CONSOLIDANTS

If a treatment is not going to last forever, should we use it in the first place? As we have seen above, we cannot be absolutely sure that the treatment will not lead to some unforeseen problem in the future. At what point should we take the risk of applying a treatment to an important stone object/monument?

Conservators paid homage for a long time to the principle of reversibility: no treatment should be used unless it can be removed at some future date, should that prove necessary. In the context of stone conservation, however, reversibility is more idealistic than realistic. It can be extremely difficult, in practice, to remove even the most soluble of treatments. It is wiser, therefore, to assume that a treatment, once applied, cannot ever be totally removed. Succeeding generations are going to have to live with the consequences of our actions.

What should we do? Treatment is irreversible, in practice, but decay through neglect is irreversible too. The dilemma highlights the importance of preventive conservation, but there are instances where preventive conservation is not enough. Ultimately it will be necessary to reach a carefully balanced decision, taking into account all aspects of each individual case. Sometimes we will conclude that treatment is justified; at other times, we may conclude that we can safely defer treatment for the time being.

This polemic is all very well, but sadly it is often irrelevant. In many of the cases with which we are confronted, the stonework has already been treated by previous generations who were perhaps less cautious or more optimistic than we may be. Often we do not know with any certainty the identity of the treatment, and often there may have been more than one treatment. This leads us to the problems of retreatment.

RETREATMENT

Virtually all research on stone treatment is based on the assumption that the treatment is to be applied to stone that has never been treated before. It is astonishing that so little work has been done on the effects that one treatment might have on another. While we hear much about reversibility, we hear little about retreatability, even though the latter is a far more important concept in practice (Teutonico et al. 1997; Van Balen, Ercan, and Patricio 1999; Hansen et al. 2003).

Any consolidant that blocks the pores of the stone and prevents the subsequent application of another consolidant must clearly be regarded with some caution. The topic that demands research, however, is the physical and/or chemical interaction of one consolidant with another. The swelling of polymers under the influence of solvents is a well-known phenomenon, but little attention seems to have been paid to the swelling of a consolidant when a second consolidant is applied. It is possible that such swelling might cause damage to the stone—which can safely be assumed to be fragile, or the consolidant would not have been applied in the first place. Moreover, one can imagine that the second consolidant will not be deposited as a coherent layer on top of the previous treatment but will form an intermingled mixture. It is not an appealing prospect, and it certainly deserves more attention. In one example, surface over-strengthening following reconsolidation treatment has been examined by Meinhardt-Degen and Snethlage (2007) using biaxial flexural strength and modulus of elasticity as the criteria.

There is equally a need to ensure that there are no unforeseen consequences of multiple applications of maintenance coatings. For example, recent work by Moreau and others (2008) has shown that the effectiveness of a water-repellent coating is reduced if it is applied either on top of, or beneath, a quaternary ammonium biocide.

RECORDING

If we cannot preserve it forever, it is imperative that we make the best possible record of stone as it exists. Indeed, one could argue that recording should have a higher priority than preserving the stone itself—provided, of course, that we are confident that the records can, in turn, be properly maintained and curated indefinitely. This is a very big proviso, and it is one that also relates to the recording of field trials. All too often, careful records may be made of treatment applications, and they are duly lodged in filing cabinets, archives, or computer disks. But ten or twenty years later, when individuals have moved on or retired, these records can disappear without a trace; much of the benefit of the trial is lost, and later conservators have no record of what was applied in the past.

Drawing and photography still have a place in recording, but attention is turning increasingly to techniques of three-dimensional recording. Molding and casting is the traditional technique, but it is not always practicable on very delicate or undercut surfaces. It is accordingly being replaced by techniques that do not entail any physical contact with the stone surface.

Raking light photography, including PTM imaging, can provide a useful documentation of the surface texture of stone, which can be later draped over a digital elevation model (DEM). Stereophotography has been known for a long time but provides only an illusion of depth from a single viewpoint. Photogrammetry, a related technique, has been widely used for producing contoured images, but it still suffers from the drawback of a single viewpoint. Holography overcomes this problem, and its use for recording sculpture was first proposed by Asmus and others (1973). Nonetheless, the role of holography has been limited largely to the production of images that, while visually striking, do not really provide a quantitative record. In this respect, laser scanning has taken the lead.

The operation of a laser scanner to record stone weathering is described by several authors (Kleiner and Wehr 1994; Ball, Young, and Laing 2000; Warrack 2000; Bates et al. 2008; Tiano and Pardini 2008). The time it takes to scan an object depends upon its size, the desired resolution, and the surface texture (shiny surfaces, such as polished or wet stone, are challenging). The primary output of the scanner is a digital record of the three-dimensional form of the object, and this can be used in a number of ways. It may be used purely as an archival record (e.g., http//:archive.cyark.org); it may be used (in conjunction with subsequent scans) to monitor the deterioration of an object (Smith et al. 2008); and it may be used to drive a milling machine in order to produce a replica (Ahmon 2004).

One of the great advantages of the laser scanner is that it does not entail any physical contact with the object and is hence suitable for even the most delicate surfaces. It has also been developed to the point where it is now capable of providing stereoscopic color images with a resolution of about 0.025 mm, which may be viewed, sectioned, and measured. Data processing and data management are significant challenges, and data clouds of 3D points are challenging to work with, in part because the 3D laser-scanning software equivalent of Adobe Photoshop does not yet exist and thus far, 3D software tends to be complex and expensive. Creating a DEM or 3D surface out of a cloud of 3D points acquired by the laser scanner requires significant processing and manipulation. In order to monitor change over time, subsequent DEMs need to be registered in three dimensions using a number of control points, whose position is known not to have changed.

The ability to produce highly accurate replicas of decayed stonework is an attractive proposition that has been seized upon by some conservators (Larson 1992; Ahmon 2004). Original sculpture can be taken indoors to the safety of a museum, while an exact copy can be put in its place. Nonetheless, there are those who argue that it may be inappropriate to install an exact copy that will be missing features already lost from the original and that it may be preferable to re-create those features in as sympathetic a manner as possible. In any event, the costs of implementing such a process remain significant, and it has not yet become a common tool in the conservator's toolbox.

As recording technologies rapidly advance, notions about the possibility of digital preservation of monuments regularly crops up, but it behooves the conservation profession to act to prevent further damage to cultural heritage as its first priority.

Chapter 5

Heritage in Stone: Rock Art, Quarries, and Replacement Stone

Issues of stone conservation extend beyond conventional notions of masonry buildings, carved gargoyles, and beautiful stone entryways. The effort to preserve our heritage in stone includes research in the conservation of rock art, the preservation of historic quarries (both for replacement stone and as historic industrial technology), the specialized conservation needs of some ornamental stones (from colored marble to mosaic stone), and the related arts of stone decoration (stone carving, polychromy, and wall painting). Finding replacement stone is an important problem in maintaining historic structures. Often the lack of timely availability of suitable stone may result in one aspect of performance being emphasized over another (e.g., initial color match, match after weathering, durability, or compatibility with existing stone). Researchers have addressed this problem in different ways in different regions.

Reviewing research related to heritage in stone is important to enhancing cross-fertilization between these specialized areas, which often have substantial interests in common. In this chapter we will focus on the first two issues—conservation of rock art and preservation of historic quarries—since the second two—ornamental stones and stone decoration—are well covered by existing reviews (Stieber 1995; Viles 2003; Henry 2006; Pepi 2008).

ROCK ART

A simple definition of rock imagery, as rock art should be called, encompasses engravings (petroglyphs) and paintings (pictographs) on rock surfaces (Ward and Ward 1996; Whitley 2001; Whitley 2005). While Lascaux in France and Altamira in northern Spain are among the best-known sites, important rock art collections are to be found around the world, notably in the southwestern United States, Australia, South Africa, India, Scandinavia, and the Sahara.

With respect to the state of the field of rock art conservation, "there has always been a large public interest in ancient pictures painted or carved on stone, but the *archaeological* study of rock art is in its infancy."[1] The same may be said for rock art preservation. One specialist has commented that, in Australia, "reactive management is in vogue and not proactive research, conservation and planning" (Watchman 2005, 17). There are, however, signs of progress. Norway implemented an extensive program for rock art preservation, undertaking an unprecedented eight-million-dollar, ten-year project between 1998 and 2008 for the research and conservation of three hundred important sites (Bjelland and Thorseth 2002; Bakkevig 2004; Bjelland et al. 2005; Hygen 2006; Gran 2008). MacLeod (2000) provides a useful review of the complex issues surrounding rock art preservation.

Interest in the conservation of cave paintings and rock art has increased significantly since the first edition of this book in 1994. The close relationship between the conservation of building stone, wall paintings, and sculpture and the conservation of rock art panels is self evident, with research opportunities for cross-fertilization between these disciplines.

Of paramount interest is the long-term preservation of fine surface details in carved stone and paint layers on stone. Traditional building stone conservation has something to learn from rock art conservation, not least as a laboratory of long-exposed art in a range of stone types and environments. Now that more rock art can be reliably dated, questions of mutual interest such as the protective and destructive aspects of lichens on longer time scales can be considered.

Approaches to the conservation of buildings and of rock art differ in interesting ways, with a number of rock art programs making effective use of volunteers for documentation and evaluating change over time, whereas in the building conservation field, such tasks have traditionally been the province of conservation professionals. A common problem in both fields is the need for useful monitoring, with Neville Agnew suggesting, "It is incumbent on us to find ways to slow rates of deterioration, which can vary enormously. One of the things not adequately studied is the rate of deterioration of rock art" (Dean et al. 2006, 11).

Rock Art Conservation

A key problem in cultural resource management is the identification of those archaeological remains in need of immediate conservation. Traditional management of rock art has focused on keeping site locations confidential, providing visitor control at public sites, and performing documentation. Natural weathering processes and conservation interventions have often not been included in conservation management planning for rock art sites. Professional conservators specializing in rock art are few, while scientific research on rock art has tended to focus on characterization and dating issues rather than conservation.

The issue of rock art's long-term sustainability and the fact that decisions must be made when allocating scarce resources suggest the need for a method of evaluating rock art stability. In some cases, rock panels have been documented to be stable over a fifty-year period (Hoerlé 2005), while other panels have been recorded as suffering rapid decay over a period of a few years (Meiklejohn 1995, 1997). The establishment of a Rock Art Stability Index has recently been proposed, involving regular recording of surface loss (Cerveny 2005; Dorn et al. 2008) to address the need for data on rates of change and to build on earlier efforts to document rock art (Turpin et al. 1979; Bell et al. 1996; Padgett and Barthuli 1997; El-Hakim et al. 2004; Barnett et al. 2005; Chandler, Fryer, and Kniest 2005; Trinks et al. 2005; Chandler, Bryan, and Fryer 2007). A range of proxy measurements of variables thought to be related to long-term stability or damage rate, such as surface temperature, have also begun to be evaluated to better manage risks to rock art (Hoerlé 2006).

There are important differences in perception, resources, and scale when comparing traditional approaches to building stone conservation and rock art conservation. Buildings are large, freestanding, or independent objects, while rock art is wholly embedded in natural settings (Dean 2001). This has raised the question of whether conservation treatments developed for problems affecting some more-durable building stone, such as biocide treatment for lichen control, are appropriate for fragile rock art surfaces (Tratebas, Cerveny, and Dorn 2004). Methods for measuring building stone decay typically require extensive training and testing and therefore carry a relatively high cost (Fitzner and Heinrichs 2002). In contrast, rock art condition assessment is often performed by volunteers using a necessarily simplified approach (Dorn et al. 2008), due to lack of resources and the large scale of the problems.

One of the most common deterioration factors for rock art is salt weathering, often from gypsum (Charola, Weber, and Bolle 1990; Hernanz et al. 2008; Meiklejohn, Hall, and Davis 2009). Strontium isotope analysis has shown that road salt from deicing has migrated to a rock art site in Norway (Åberg, Stray, and Dahlin 1999) and has caused crystallization damage. Analysis of important rock art in Nine Mile Canyon in central Utah shows that the use of magnesium chloride salt as a treatment to reduce dust on dirt roads through the site appears to be having a deleterious effect on adjacent rock art (Kloor 2008). The dust raised by truck traffic on the road is also obscuring the visibility of some petroglyphs and paintings. Road dust has also been found to obscure rock art at other sites as well (Watchman 1998).

The extremely well-preserved condition of cave paintings at Lascaux, Altamira, and elsewhere led to the realization of the critical role microbes play in the long-term stability of cave paintings. Rapid cave painting decay following disturbance of the microbiological environment has reminded conservators that our knowledge of microbial decay is still inadequate (Dornieden, Gorbushina, and Krumbein 2000). The debate over how to respond to microbial "outbreaks" on cave paintings has spilled over into the popular press (Allemand and Bahn 2005; Castellani 2005; Pringle 2008; Bahn 2008). Research into microclimate stabilization and shelters for rock art sites has found that such systems can significantly improve environmental stability and that human visitation often negatively affects the stability at cave sites (Dragovich 1981; Wainwright, Sears, and Michalski 1997; Hoyos et al. 1998; Brunet, Vouvé, and Malaurent 2000; MacLeod and Haydock 2002; Sanchez-Moral et al. 2005; Canals i Salomó et al. 2005; Brunet, Malaurent, and Lastennet 2006).

Progress in our understanding of how microbial activity can damage rock art has improved over the past decade. For example, MacLeod and others (1995) have documented an increase in surface acidity related to increased seasonal moisture on Aboriginal rock art surfaces. Rapid lichen growth over rock art in Australia was found to be related to the amount of sunlight falling on rock surfaces, resulting in proposals for shelters and other minimally invasive lichen-control methods (Ford and Officer 2005). In South Africa, cracks in pigment layers are allowing water and fungi to penetrate rock paintings (Arocena, Hall, and Meiklejohn 2008). In an example of preventive conservation, researchers caution against removing any vegetation that provides thermal buffering of rock art surfaces (Hall, Meiklejohn, and Arocena 2007).

In Norway, a series of proposals have been made to reduce the rates of damage to rock art, including sheltering, reburial, and modification of environmental conditions to help neutralize acids and reduce oxidation (Walderhaug 1998). A review of decay mechanisms at these sites found frost and tree roots to be of greatest concern, with acid rain and mineral leaching of lesser importance (Walderhaug and Walderhaug 1998). The use of insulating materials on Scandinavian rock art was found to significantly reduce the impact of freeze/thaw cycles (Gran 2008). A recent dissertation on lichen damage to rock art offers advice to heritage managers (Dandridge 2006).

Fire has long been recognized as a deterioration factor for rock art. Research shows its effects are more widespread than previously thought, and preventive measures, such as clearing vegetation by hand, are recommended to reduce fire risk (Tratebas, Cerveny, and Dorn 2004).

Similar problems have been found in different parts of the world. For example, conservation assessments of rock art sites in Bolivia found damage from graffiti, salts, humidity cycling, and uncontrolled tourism and proposed more integrated site management (Taboada Téllez 2007). A case study in Brazil similarly found salt efflorescence, dust, and animal activity (nests and droppings) resulted in fading, flaking, and loss of readability of rock art panels (De Oliveira Castello Branco and Cruz Souza 2002). A higher level of coordination and information sharing in rock art conservation research would be beneficial.

Rock Art Treatment

Conservation treatments to date have included moisture control, consolidation of rocks and pigments, removal of mud nests and lichens, graffiti removal, surface cleaning, and repair of scratches or gunshot damage resulting from recreational firearm use (Pearson and Clarke 1978; Andersson 1986; Lambert 1988; Rosenfeld 1988; Brunet, Guillamet, and Plassard 1997; Dean 1997; Dean 2001; Jeyaraj 2004). In a recent treatment example, test areas of schist in the Côa Valley of Portugal were treated to evaluate the long-term effects of drainage and flood protection. Outcrops were covered with "reinforced soil," and openings between blocks were filled with layers to encourage drainage and normalize the surface (Batarda-Fernandes and Delgado Rodrigues 2008). In another example, a range of biocide treatments for algae on marble petroglyphs were tested, and several were found to be effective (Laver and Wainwright 1995; Young and Wainwright 1995).

A small core of specialist conservators has worked in the field of documentation and conservation of rock art sites (Dean 1998; Whitley 2006). An overview of treatments used on rock art is the subject of a master's thesis (Dandridge 2000). The use of organic glue to stabilize fragments and cement mortar to fill cracks in rock art panels has been evaluated and criticized (Bakkevig 2004). Researchers have tested antifungal and antibacteriological treatments to help mitigate biodeterioration of rock art (Gorbushina et al. 2003). The EC has sponsored research into rock art conservation, including a project titled "Non-destructive technique for the assessment of the deterioration processes of prehistoric rock art in karstic caves: The paleolithic paintings of Altamira" (Zezza 2002).

The final report on Norway's ten-year rock art preservation program concluded with an appeal to restrict the use of Mowilith[2] DM 123 S for conserving rock carvings, due to stability problems and the fact that Mowilith swells with the addition of ethanol. In Norway, ethanol is used to remove lichens, so Mowilith is now considered an incompatible material for which the long-term effects are unknown (Hygen 2006). Over the evolution of the project, the primary investigator became more reserved concerning direct interventions, and the project moved increasingly in the direction of indirect and preventive methods. Norwegian research has also advanced the debate over whether lichens are protective, neutral, or damaging to rock art, finding some lichens are more aggressive than others (Bjelland and Thorseth 2002; Bjelland and Ekman 2005; Bjelland et al. 2005). However, current Norwegian guidelines discourage lichen removal through chemical treatment and indicate that "removal of lichens should only be done in the instances where there are binding plans for regular follow-up of the actions" (Hygen 2006, 19).

Approaches to cleaning of rock art sites also vary according to the specific environmental problems affecting the art. To remove graffiti at the cave of Rouffignac, the ceiling was cleaned with compresses soaked in a diluted ammonia solution, while in areas where the graffiti was more difficult to remove, special erasers of differing density were used (Brunet, Guillamet, and Plassard 1997). In Zimbabwe, paint stripper and toluene were recommended to clean the graffiti from rock art (Taruvinga 2003), because, when tested, laser cleaning was found to remove both the graffiti and the original paint layers.

The overall impression gained from this literature survey is that the need to protect rock art has led to treatments being applied somewhat ahead of the scientific study of appropriate interventions. Nonetheless, looking beyond the material decay of rock art to the problem of increasing tourism and the need to link rock art conservation efforts to local economic development has received some needed attention in recent years (Walderhaug Saetersdal 2000; Smith 2006; Deacon 2006).

Rock Art Documentation

Given the large number of rock art images and sites worldwide, documentation is seen as an important preservation tool. In a number of countries, documentation of rock art has been largely implemented by volunteers (Chandler, Bryan, and Fryer 2007). A survey of those involved in documenting UK rock art sites found a perception of rapid, variable degradation from the impact of humans and animals, superimposed over a slow background level of erosion caused by physical and chemical agents (Barnett and Díaz-Andreu 2005). Laser scanning of rock art as a way to monitor decay rate has been researched by several authors (El Hakim et al. 2004; Barnett et al. 2005; Trinks et al. 2005). One of the reasons for making rubbings of rock art and gravestones is that fine details not visible to the naked eye can be recorded using this method. An attempt to record an example of fine detail using 3D laser scanning was undertaken (Díaz-Andreu et al. 2006), but scanning was not able to detect a spiral feature recorded in a rubbing in 1995.

Good documentation has led to preventive conservation recommendations, such as the employment of mitigation practices to reduce the abrasive effects of blowing sand (Keyser, Greer, and Greer 2005). Knowing that most of the four hundred thousand rock art sites around the world (Clottes 2006) will never receive a conservation intervention, let alone a conservation assessment, good documentation (often by volunteers or students) has been the most common method of capturing a durable and accessible record of these sites (Padgett and Barthuli 1997; Swartz and Hale 2000; Larkin 2002).

HISTORIC QUARRIES

The preservation of historic quarries is of interest to the field of stone conservation for several reasons (Ashurst 2007, 306). One is that quarries provide important evidence of how stone production technology has evolved. This technology has a profound effect on stone durability, as we saw with the issue of the bowing found in thin marble panels. The thinner panels were made possible by new production technology (Scheffler 2001). Second, historic quarries may need to be reopened to provide replacement stone for important buildings. In Sydney, the local "yellow block" sandstone is being quarried when the foundations of modern skyscrapers are dug and stockpiled by the local conservation authorities for later use as replacement stone on nearby historic buildings.

Perhaps the most extensive work on ancient quarries is the EC-sponsored project known as QuarryScapes (Conservation of Ancient Stone Quarry Landscapes in the Eastern Mediterranean), coordinated by the Geological Survey of Norway (2005–8), which dealt with issues of inventorying, managing, and conserving ancient quarry sites, with case studies in Egypt, Jordan, and Turkey (Abu-Jaber, Al Saad, and Al Qudah 2006; Bloxam 2006; Degryse et al. 2006; Heldal et al. 2006; Caner Saltik 2007; Heldal, Bloxam, and Storemyr 2007). The other main resource for information on historic quarries is the research group ASMOSIA (Association for the Study of Marble and Other Stones in Antiquity; ASMOSIA.org). A series of conference proceedings contains the publications of geologists and archaeologists working on discovering ancient sources of stone, as well as stone transport, trade, conservation, and archaeometry (Schvoerer 1999; Lazzarini 2002; Herrmann, Herz, and Newman 2002). Another useful volume on Egyptian quarries is the work by Klemm and Klemm (2008).

REPLACEMENT STONE

Related to the preservation of historic quarries is the issue of obtaining adequate replacement stone for repairs, which has become a critical problem for many important sites as historic quarries are closed due to development and other economic pressures. Standards and resources for replacement stone vary enormously from country to country, and a critical review article on this topic with a global view is overdue.

When repairs are being planned for a large building, quarrymen and geologists are often asked: "Which of the available stones will provide good durability and a compatible match to the existing stone?" (Jefferson et al. 2006). During the renovations of the British Museum, the "wrong" stone was used (Niesewand 1999). Recent research by Rozenbaum and others (2008, 345) found that for French limestones it was difficult, but not impossible, to "select substitution stones with satisfactory aesthetic aspect and properties that enable to expect a satisfactory compatibility with the original stone."

Finding appropriate replacement stone requires tools for stone selection such as atlases and databases (Dingelstadt et al. 2000; Hyslop et al. 2009). A useful discussion of aspects of selecting replacement stone based on material properties can be found in two recent works (Přikryl 2007; Yilmaz 2008).

Clearly, to improve on the current situation, each country with significant heritage in stone should have a centralized lithological library, and an associated database, that includes not only the petrographic and mineralogical characteristics of its stone but also petrophysical ones, including pore size distribution, porosity, capillary uptake coefficient, and hydric and hygric dilatation.

To this end, English Heritage is working with the British Geological Survey and local experts to expand their database of English stone with a new GIS site called EBSPits (England's Building Stone Pits) and to identify the most important building stones used, representative buildings, and historic quarries (English Stone Forum 2009; English Heritage 2009).

Research by Blanc and others (Blanc and Lorenz 1988; Blanc and Lorenz 1992; Holmes, Harbottle, and Blanc 1994) has helped to identify many quarries in France, in part with the goal of architects being able to better match replacement stone. The Bureau de Recherche Géologique et Minière (BRGM) and the Laboratoire de Recherche des Monuments Historiques (LRMH) are collaborating on a project to gather into a database all the information related to the stones of monuments, ancient quarries, and modern quarries (V. Vergès-Belmin, personal communication).

Recent work by Hyslop and others (Hyslop and McMillan 2004; Hyslop 2008) discusses the challenges of finding replacement stone for the important and well-known stone buildings of Glasgow and Edinburgh. In Glasgow, Duthie and others (2008) found significant variations in the extent of microbial growth on a range of replacement sandstone blocks that had been exposed for twelve years, illustrating the importance of selecting appropriate replacement stone.

Related to the issues of stone replacement is the general question of loss compensation for stone. This topic has been reviewed by Griswold and Uricheck (1998), who suggest that this area be prioritized in future research and evaluation.

Notes

1 Whitley 2001, book jacket.

2 Mowilith is an aqueous emulsion of polyvinylchloride, polyvinylacetate, and different stabilizing agents.

Chapter 6

Doing Better: Increasing the Effectiveness of Research

WHAT IS WRONG?

The purpose of this chapter is to suggest some ways in which our research might be made more effective. The views expressed are unashamedly personal, and not everyone will agree with them. It is hoped, nonetheless, that they will stimulate some serious thought and discussion in order that limited research resources may be put to the best possible use.

In the last fifteen years, three factors have helped increase the effectiveness of research: the entry of topflight researchers into the field, increased access to existing research via Internet databases and PDFs, and the increase in research done by universities, particularly those participating in EC-sponsored programs. Much work in conservation and conservation research is published only as "gray literature," and the Internet has vastly increased accessibility to this material (see, for example: http://repository.upenn.edu/hp_theses; http://www.ncptt.nps.gov/product-catalog/). There have been some corresponding changes that have decreased the effectiveness of research over this period as well: the decrease in funding of research programs at the institutional and national level (BRE, CSIRO, GCI, ICCROM, national research programs, etc.),[1] the need to test university innovations and transfer them to the field, and the need for longer-term research programs. These factors are discussed in more detail below and in the final chapter.

Publications

The number of published papers relating to stone is growing relentlessly. Every four years a large stone conservation meeting is held, and this is reflected in the overall pattern of publication.

On the face of it, this must surely be welcomed. It indicates the growing concern about stone and the growing numbers of researchers who are working on stone, many of whom bring important new perspectives and discoveries. However, the quality of many of the papers is still disappointing. Why?

The following criticisms are often made:

- The same material is published on more than one occasion. While it is acceptable to publish interim reports on a major piece of work, there is no excuse for publishing the same material, with only minor variations, time and time again.

- Many papers consist of the application of well-tried procedures to a specific building or monument. The results are of interest only to a limited audience and should be written up as an internal report of the organization carrying out the research. They do not warrant full publication in journals or conference proceedings.
- Many papers fail to set the research into context. They are essentially descriptive; they describe the work that was undertaken but do not say why it was done.
- Many papers neglect to indicate the significance of the results. Having failed to say why the work was done, they provide insufficient discussion of the results and therefore do not explain what, if anything, was achieved. The reader is left wondering whether any advance was made and, if so, what it was.
- Few papers identify promising avenues for further research.
- Underlying the previous problems is the frequent neglect of the scientific rigor of hypothesis—experiment—conclusion.

Conferences

Conferences provide unparalleled opportunities for meeting fellow researchers: for making new contacts, finding new collaborators, comparing notes, sharing ideas, and keeping up to date. They also provide a much-needed opportunity to stop and think and to see one's research in a broader context.

On the negative side, however, conferences often provide an opportunity for publishing substandard, nonrefereed work. The proliferation of conferences, however desirable it may be, can all too easily lead to a proliferation of poor-quality papers. These and related issues have recently been addressed by the Torun Guidelines for stone meetings, which serve as an example of what can be done to improve stone conferences (see sidebar, page 68).

Standards

The lack of internationally agreed-upon standards, be they for nomenclature or for testing procedures, hinders the interpretation, understanding, and evaluation of research. Without standards, there is no common language. The situation is slowly improving, with the adoption of English as the current language of science, which provides greater opportunities for communication and collaboration among researchers and research groups, and with collaborative tools and more universal evaluation standards beginning to be adopted (Fassina 2008; European Committee for Standardization = Comité Européen de Normalisation), such as drilling resistance and ultrasonic testing.

Conduct and Quality of Research

Research into stone conservation demands an interdisciplinary approach. Many researchers, however, find themselves working alone or in relatively

The Torun Guidelines for Conferences in the Field of Stone Conservation

Introduction

In an era of increasing information and changing dissemination technology it seems an appropriate moment to reflect on ways to improve the quality and accessibility of knowledge in the field of stone conservation.

As knowledge increases rapidly, teams working on stone conservation have become more specialised and often present their results at specialist meetings. This trend may increase the potential for isolated perspectives and the risk that knowledge may not reach its intended goals.

The general congresses on stone deterioration and conservation, organised every 4 years since 1972 give a useful snapshot of the different trends of stone conservation and provide a multidisciplinary forum for discussion, complementing the specialist meetings. However, it can be difficult for them to encompass all the different trends and fields of stone conservation.

In recent decades there have been a number of calls to improve the quality and impact of knowledge in the conservation field. In response, there have been a number of improvements, such as more review articles and multi-author textbooks which give new researchers some of the background needed. Electronic publication of full text articles from most journals makes the peer-reviewed literature more readily available. Nevertheless, most conference proceedings still have limited electronic distribution.

With the aim of improving the quality and the dissemination of knowledge through congresses in the field of stone conservation, the 11th International Congress on Deterioration and Conservation of Stone, and the 13th meeting of the ICOMOS International Stone Committee, which met in Torun on September 15th to 20th 2008, adopted the following text.

The Guidelines

1 Planning

When planning conferences organisers should review other conferences already scheduled in the field, in order to separate their own conference from others by at least six months. The aim is to increase the potential pool of participants and to increase the likelihood of original research being presented.

2 Selection of papers

The selection of papers for formal conferences should be based on a thorough review by at least two experts. Organisers, assisted by their scientific committees, should check for and refuse 'doublons,' i.e. papers that have been, or are about to be, published in proceedings of another conference. Published papers (whether oral or poster) should meet minimum standards, including:

- precisely defined research methodologies
- appropriate reference citations
- advancing knowledge in the field.

3 Communication among participants

Organisers should encourage formal and informal communication among conference participants. These may include discussion sessions, panel discussions and workshops.

4 Seeking quality and measuring outcomes

Organisers, assisted by their scientific committees, should ensure good quality papers. In addition, organisers should measure the outcomes of their conference. Measures adopted may include reviews of the conference and opportunities for user feedback, such as a web page for participant responses, and quality rankings.

5 Dissemination strategy

To facilitate rapid dissemination of the ideas presented at the conference, organisers should plan for electronic dissemination of the proceedings. This should be arranged within a short period of time (e.g. a year) to ensure that the results achieve a wide and long-lasting distribution.

The following persons participated to the drafting of the Torun Guidelines:
Akos Török, Hungary — Clifford Price, UK — Dagmar Michoïnova, Czech Republic — Daniel Kwiatkowski, Sweden — David Young, Australia — Elsa Bourguignon, France — Eric Doehne, USA — Hilde De Clercq, Belgium — Jadwiga W. Lukaszewicz, Poland — Jean-Marc Vallet, France — Jo-Ann Cassar, Malta — Johannes Weber, Austria — Jose Delgado Rodrigues, Portugal — Milos Drdacky, Czech Republic — Marisa Laurenzi Tabasso, Italy — Myrsini Varti-Matarangas, Greece — Philippe Bromblet, France — Stefan Simon, Germany — Vasco Fassina, Italy — Vasu Poshyanandana, Thailand — Véronique Vergès-Belmin, France.

http://www.iccrom.org/eng/news_en/2009_en/field_en/01_01TorunGuidelines_en.pdf.

small teams. As a result, research can become too narrow, failing to take into account factors that might seem self-evident to somebody trained in another discipline. For example, an analytical chemist might look primarily at the composition of a stone, while a materials scientist would perhaps focus on its behavior, a biologist could discover a new species of microbe in the pores, an engineer would core the stone and measure its strength, and a geologist might make a thin section and evaluate its microtexture.

This recalls the famous story from India of the truth being similar to a group of blind men trying to describe an elephant by touching it, when each has access to just a single, and different, part of the beast (trunk, leg, ear). A corollary of this situation is the phrase: "When the only tool you have is a hammer, it is tempting to treat everything as if it were a nail" (Maslow 2006, 15). It is useful for researchers to work closely with conservators and conservation architects to mitigate such tendencies. At worst, researchers can become so introspective that they take little or no account of work being undertaken elsewhere; the researcher whose citations are solely to his or her own work is clearly falling into this trap.

A great deal of research into stone is conducted at a rather superficial level, but this is changing. The first volume of this book complained that much of the work on consolidants, for example, was very empirical, and that a particular material would be evaluated simply because it was available, not because there were sound theoretical reasons for believing that it would be effective. Some work on decay mechanisms was seen as equally superficial. However, the depth of research has increased enormously in the intervening years, and some areas have changed beyond recognition, thanks to the contributions of exceptionally talented individuals. Nonetheless, there is still a danger that research can become so theoretical that it loses sight of its main purpose. The researcher needs to be fully aware of what is desirable and practicable from a conservation standpoint, while conducting research at a level that is deep enough to solve the fundamental problems.

Some of these issues were summarized succinctly by Chamay (1992) in his closing remarks at a conference:

> *Je m'inquiète un peu de constater que vos recherches sont menées sans concertation organisée, chacun travaillant de son côté, l'échange d'information restant très limité . . . J'ai aussi le sentiment que la tendance générale parmi les chercheurs est de rester confiné dans sa spécialité . . . Attention à l'arbre qui cache la forêt! Avant d'entrer dans le détail, une appréciation d'ensemble est nécessaire.* [I am a bit worried to notice that you are carrying out your research without organized dialogue, each person working in his or her own corner, the exchange of information remaining very limited . . . I also have the feeling that the general tendency among researchers is to remain confined to one's own specialty . . . Don't fail to see the wood for the trees! Before going into detail, an assessment of the whole is necessary.]

It is interesting that one of the concerns that led to a recent conference on the conservation of the cave at Lascaux (AP 2009) was the need to have specialists working more closely together and synthetically, just as Chamay suggested.

Getting the Message Across

There is no point in doing research unless the outcome can be applied in practice. This does not mean that there is no place for long-term, strategic research, but that any worthwhile research must ultimately contribute to the care and conservation of the heritage.

There are many ways of getting the message across, including lectures, publications, personal contacts, and advice on specific problems. The message needs to reach other researchers, but it must also reach, for example, conservators, architects, archaeologists, and administrators. It does not follow automatically that a good researcher is a good communicator, and all researchers should ask themselves whether their research is achieving the full impact it deserves. The Internet has changed expectations about the ease of access to high-quality information.

PUTTING IT RIGHT

What can be done to make our research more effective? There are no simple solutions. While some steps may be taken by individual researchers, other solutions lie with research administrators, conference organizers, editors, publishers, training institutions, and funding bodies. The following proposals deserve consideration.

Quality, Not Quantity

Any institution that funds research may reasonably expect to see some return for its money. This necessitates some means of measuring research output. How else may the institution be sure that its money is being well spent? The simplest indicator, and one that appeals to many administrators, is the number of papers that result from the research. It is an objective, quantitative indicator, but it is one that undermines quality. Individuals find themselves under immense pressure to produce a certain number of publications each year, and it is no wonder that quality suffers. Publishing the same thing several times is an easy way of meeting the target. Other tactics include the publication of a string of interim reports, the publication of material that warrants no more than an internal report, publishing papers that report on what one proposes to do in the future, and publishing papers with a long and unjustified string of authors.

Journal impact factors and citation indices, such as Google Scholar, ISI Web of Knowledge (Science Citation Index), and Scopus (by Elsevier), can provide an indication as to what references and journals are having the most effect. The number of times an article is cited is tracked as a measure of its popularity and potential usefulness. Journals that contain articles that have higher rates of citation have higher impact factors.

The use of citation indices is rapidly increasing in biology and medicine as an important way to filter the wheat from the chaff of research publications and to provide employers with an independent assessment of quality, such as the H-index (http://en.wikipedia.org/wiki/H-index). As with all rankings, the system is open to abuse. Popularity is not the same as quality, since once an article begins to be cited, its chances of being cited again increases. Assessment of quality in research is not a simple matter of numbers. It entails a high degree of subjective judgment, both by research managers and by other researchers. Funding bodies must be prepared to appoint research managers whose judgment they trust and then be prepared to accept that judgment concerning the quality of research being conducted under those managers' supervision. They must be seeking value for money, which entails both quality and quantity, rather than quantity alone.

Conferences and Other Models for Advancing the Field

Other types of scientific meetings should be considered as role models, such as workshops, the Gordon Research Conferences (GRC) (2010),[2] the Dahlem Conferences (Freie Universität, Berlin 2006),[3] and other forums based on new technologies, such as online discussions of presentations at conferences and online proceedings where attendees can post questions and comment on articles. Conference organizers today can take for granted that most participants possess a Wi-Fi–enabled laptop, netbook, or cell phone. Meetings where participants can actually take part in discussions, interact, and comment in concrete ways are often more productive than conventional meetings at which participants are often overwhelmed by too many presentations, too much information, and too little time for useful discussion.

Conference Papers

It is a common practice for employers not to fund an individual's attendance at a conference unless he or she is presenting a paper or a poster. It is a practice that makes the research administrator's life much simpler, but one that again encourages the production of superfluous publications. To solve this problem, one option would be amending the conference attendance policy to include publishing a conference review as a qualifying activity. At a large conference, multiple reviewers, who may be paid small honoraria for their contribution, can cover parallel sessions. A good conference review is often worth more than several case studies.

Selection of Conference Papers

Another important way of preventing the publication of substandard conference papers lies with the conference's technical committee. All too often, papers are selected on the basis of an abstract submitted some eighteen months or more before the conference. At that time, the research will almost certainly not have been completed; indeed, it may not even have begun. The prospective author, therefore, makes a guess as to the likely outcome of the research and writes an abstract that strikes a delicate balance between the specific and the noncommittal. The technical committee reviews the abstracts and, on this flimsy evidence, decides which papers to accept. By the time acceptance is gained, the author has twelve months or less in which to complete a paper—regardless of how the research is going. Then, the technical committee and the editors, when they finally receive the paper, have little option but to publish it much as it stands.

Not all conferences operate this way, but many do. It means that much of the literature of conservation has been subjected to the very minimum of refereeing, if any. Quality assurance is all but nonexistent.

If preprints are to be issued at the time of the conference, there may be insufficient time for full refereeing. Nonetheless, a significant step forward could be made if technical committees were to insist on seeing the full text of a paper before deciding whether to accept it for presentation and publication. It is true that it takes longer to read a paper than it takes to read an abstract and that technical committees are composed of busy people. However, it does not take long to decide whether a paper consists largely of previously published material or whether it is of local interest only. A lot of substandard papers could be weeded out very quickly. Another problem is that authors may not be prepared to take the time to write a full paper if there is a risk that it may not be accepted. Too bad—if poor papers were weeded out, there would be correspondingly more space for good papers, so the author who has something worthwhile to say need not fear rejection.

Refereeing

Ideally, all published material should be subjected to peer review. It is a process that is open to criticism in that it slows down publication and can fall afoul of an ill-informed or prejudiced referee. It is, however, the fairest way of ensuring that papers are of sufficient quality to merit publication. Conservation has suffered greatly from the fact that so much of its literature has been in unrefereed publications. As one conservation scientist observed: "Why try harder, when you can get away with being sloppy?"

Collaborative Programs

The time has long passed when a well-educated individual might have a working knowledge of the whole of science and the humanities. We are all highly specialized in our individual fields, and we need to collaborate with specialists in other disciplines if we are to solve the very broad problems posed by stone conservation. Not only do we need to collaborate with other conservation scientists from different disciplines, but we also need to draw in talented researchers who are not involved in conservation. Such collaboration is not without dangers (Torraca 1999), but it is essential nonetheless.

Some funding bodies are in a position to enforce collaboration. An example is to be found in EC-operated programs. Research projects are not funded unless they entail genuine collaboration between partners in more than one member state, with each partner making a clearly defined contribution based on a particular expertise. In a relatively short time, these programs have brought about a much greater degree of collaboration between relevant European research institutions.

Training

Good research requires good researchers. To be a good researcher in the scientific aspects of stone conservation, one needs a thorough grounding in science, training in research, and a sound appreciation of conservation issues. These qualifications are not readily found in any one individual, and a significant proportion of "conservation scientists" do not have sufficient knowledge of science to enable them to undertake research at a fundamental level. They may, for example, have trained primarily as conservators; although their training may well have included some science, they are conservators first and scientists second. As a result, a good deal of research is rather superficial.

Much attention has been paid to the training of conservators, and lists of training courses are readily available (Rockwell 1994) (also see appendix, List of Conservation Related Sites, pages 150–51). Less attention has been paid to the training of conservation scientists, although there has been some useful discussion of the different approaches to training researchers (Mazzeo and Eshøj 2002; Chiari and Leona 2005; Trentelman 2005; Mazzeo and Eshøj 2008). There are very few training programs for conservation scientists (http://www.episcon.scienze.unibo.it), and a worldwide survey of current training opportunities would be advantageous. A number of possible pathways can be envisaged: doctoral research followed by a fellowship in a major conservation institution or museum, for example, or a master's degree in a particular aspect of conservation science. A first degree in a scientific subject should be a prerequisite, in any event.

Some attention also needs to be given to ways of attracting high-caliber students to conservation. The subject does not, on the whole, attract the outstandingly capable researcher. Such individuals are more likely to be found in medical research, in nuclear physics, or in military research, where they will benefit from better funding and from the stimulus of working in large, highly focused teams. Ways must be found of bringing conservation to the attention of science students during the course of their first degree and of presenting stimulating and challenging career opportunities. Part of this issue is that to be more successful, the conservation field needs to scale up its ambitions and build support for larger-scale, coordinated projects to compete with "Big Science." In the United States the Mellon Foundation has been effective in bringing scientists into museums through endowed chairs;[4] however, permanent positions dedicated to monuments research remain unconscionably rare worldwide—an important gap that could be filled with the requisite institutional support.

Reviews

The conservation literature is still remarkable for its relative lack of scholarly review articles. In all the mainstream scientific disciplines, the need for state-of-the-art reviews is well recognized, and the authors highly acclaimed. The conservation literature, by contrast, is full of isolated pieces of work, with very little effort being made to pull the information together.

Review articles enable researchers to put their work in context and to see where further work would be worthwhile. However, they are not easy to write. They require a lot of time and a high degree of competence. They may have to be specifically commissioned and funded. The National Center for Preservation Technology and Training (NCPTT) has funded some small grants for researchers to write reviews, and the International Institute for Conservation (IIC) journal *Reviews in Conservation* has proven itself to be an extremely useful resource to the conservation community.[5] Progress is certainly being made, but there is plenty of scope for more.

Notes

1 BRE = British Research Establishment; CSIRO = Commonwealth Scientific and Industrial Research Organization, Australia; GCI = Getty Conservation Institute; ICCROM = International Centre for the Study of the Preservation and Restoration of Cultural Property, Rome (UNESCO).

2 Gordon Conferences are organized around a theme, with few presentations, much discussion, and with contributions "off-record" to encourage free exchange, often of unpublished material.

3 "The Dahlem Conference in Berlin is a unique forum for analyzing, in a multidisciplinary way, complex topics. For five days fifty selected participants are cloistered together, divided into four groups. Each group studies background papers prepared by a few selected individuals, which serve as a basis for further discussion and the preparation of a report. The goal of these reports is to define what is *not* known in the field rather than rehearsing what is known and to present ideas for further research and their priorities. The four group reports are hammered together under the guidance of a rapporteur" (Wolff 1992).

4 Museum science involves research and service work in support of curators and conservators and revolves largely around issues of technical art history as well as art conservation.

5 Regrettably, the recent financial crisis, or "great recession," has led to the consolidation of the journal *Reviews in Conservation* into the IIC journal *Studies in Conservation*.

Chapter 7

What Has Changed? Some Thoughts on the Past Fifteen Years

The first edition of this book was written at a time when research on stone seemed to many people to have stagnated. That perception has changed completely in the intervening years, and real progress has been made in many areas. Significant gaps in knowledge have been substantially narrowed, including many of the fundamental aspects of damage to stone from cycles of frost, salt, moisture, and heat. In a number of cases our descriptive nineteenth-century notions of "weathering" have now been deeply probed and quantified, measured in the field, and replicated in laboratory experiments. These insights have in some cases led to innovations in the preventive treatment of stone, with more sophisticated models of damage advancing hand in hand with more quantitative observations from field measurements and laboratory experiments.

Five important trends can be identified: 1) a perception that our understanding of fundamental conservation problems is far ahead of solutions to these problems; 2) new solutions to stone conservation problems often need long-term testing, but resources for such testing are lacking; 3) climate change is an important issue in stone conservation; 4) biodeterioration should be increasingly understood in an ecological context; and 5) the locus of stone conservation research activity may be beginning to shift to countries such as China, India, Brazil, and South Korea. Heritage conservation in these countries is becoming a national priority due to unacceptable rates of heritage loss and greater economic success.

The traditional neat classification of weathering mechanisms into physical, chemical, and biological factors is receding as an accepted approach to this field. A new approach emphasizes material behavior and the important interrelationships between environmental, material, and historical variables. As is the case with most natural systems, a few key parameters often dominate in each weathering process (Goudie 1995), and the result can be nonlinear and even chaotic, in contrast to previous assumptions about linear rates of erosion. This schism is reminiscent of the nineteenth-century debates over catastrophism and Darwinian gradualism (Viles 2005; Giavarini et al. 2008).

The straightforward concepts of magnitude, frequency, and dose-response developed for air pollution studies on stone (Charola and Ware 2002) are being modified by ideas of thresholds, feedback loops, and nonlinearities (Goudie and Viles 1999; Norwick and Dexter 2002) within the large framework of conservation risk assessment (Brokerhof et al. 2007).

One example of this profound shift over the past two decades is the conservation of the wall paintings of Queen Nefertari since the late 1980s. Development of the conservation program went through several steps: 1) assessment, monitoring, and conservation of the wall paintings in the late 1980s; 2) visitor impact (carrying capacity) studies in the 1990s; and, 3) most recently, research that suggested that the greatest risk to the wall paintings appears not to be humidity cycles that might activate salt weathering but instead rare flash floods in the area. Accordingly, the need to prepare for rare, but catastrophic risk has assumed the same importance as the need to manage more gradual risk factors (Agnew and Maekawa 1999; Wüst and Schlüchter 2000; McLane et al. 2003).

In one of the most important trends over the past fifteen years, universities have embraced many of the compelling multidisciplinary challenges found in stone conservation, bringing to bear new, topflight researchers and new tools from materials science, cement chemistry, geotechnical engineering, geology, physics, geochemistry, microbiology, and geomorphology, and adding these to the historic tradition of chemists at the center of much research in stone conservation. As a consequence, the standard of research has improved beyond recognition in some areas, and many more papers are being published in the peer-reviewed mainstream literature. The publication of an increasing number of reviews is also much to be welcomed.

One discipline that should be added to the mix is that of (for want of a better term) heritage hydrology. This is the nanometer- to kilometer-scale study of the effects of water transport on the stability of historic architecture and monumental complexes. Archaeologists have long realized that hydrology was critically important in sustaining the cultures that built Tiwanaku, Copán, Moenjadaro, Baghdad, Petra, and Angkor, for example. And for architects, one of the key elements of building design is how a structure sheds water. Increasingly sophisticated models of moisture transport and material behavior are developing rapidly (such as WUFI, ASTRA, or CESA) (Sedlbauer and Künzel 2000; Holm and Künzel 2003; Franke et al. 2007). These areas have also benefited from advances in the modeling of the structural behavior of building materials (Binda 2007). Future researchers in monument conservation should be encouraged to specialize in heritage hydrology.

Multidisciplinary research has been strongly promoted by EC-funded research projects over the past fifteen years, resulting in an experienced network of about eighty multidisciplinary researchers across Europe with an interest in the subject. Indeed, it could be argued that the gradual evolution of this informal research network has been of even more value in promoting research than the projects themselves. Monuments research networks supported by similar levels of funding are currently lacking in the Americas and Asia.

It is uncertain whether overall research funding has improved or declined—there do not appear to have been any attempts to gather the necessary data. There is, however, a general impression that expenditure on stone research has increased somewhat within the university sector and declined substantially elsewhere. For example, expenditure by governments and NGOs, such as BRE, CSIRO, EH, GCI, ICCROM, NPS,

NCPTT, TNO, and the former Swiss Expert Centers has generally decreased over the past fifteen years.[1]

In the United States, research funding for stone conservation remains difficult to obtain, with some work on biodeterioration and building materials receiving limited National Science Foundation (NSF) support and more applied research funding coming in the form of small grants from the National Center for Preservation Technology and Training (NCPTT) and the Kress Foundation. A July 2009 Mellon Foundation–sponsored meeting with the US National Science Foundation on heritage and science suggests that there is growing interest in the field at the national level in the United States. However, one colleague has quipped, "The larger science community 'rediscovers' conservation about every ten years. But when they find out that the problems are difficult and that funding is scarce, they lose interest."

In an encouraging development, funding is coming increasingly from outside Europe and the United States. Researchers in India and China are beginning to publish conservation research at a greater rate, partly in response to significant challenges from air pollution, tourism, and climate change. Russia, Brazil, and South Korea also have seen growing interest in research related to conserving heritage in stone as their economies have developed.

Despite the vagaries of funding, the number of research publications has continued to increase, as researchers in allied disciplines, from geography to materials science, have discovered compelling scientific challenges in the field.

Stone conservation issues have increasingly been covered in the popular press, including public policy as regards conservation research (House of Lords, Science and Technology Committee 2006; House of Lords, Science and Technology Committee 2007), the biodeterioration of monuments (Venkataraman 2008), and the crumbling of cathedrals (Petre 2006). The sites of Easter Island, Petra, and Angkor are compelling for conservation professionals and the public alike, not only for their beauty and history but also because they are inexorably eroding as unresolved conservation and funding issues continue. Petra (Paradise 2005; Simon, Shaer, and Kaiser 2006; Heinrichs 2008) and Angkor (Leisen 2002; Leisen, von Plehwe-Leisen, and Warrack 2004; André et al. 2008; Siedel, von Plehwe-Leisen, and Leisen 2008) are also important examples of new knowledge from conservation research being brought to bear.

The increasing importance of English as the current common language at most conferences and for many journals has helped to consolidate the field of conservation.[2] Nevertheless, the fact that many important works of research are published only in the French, German (see appendix), and Italian literature remains a barrier (Cabreroravel 1993; Alessandrini and Pasetti 2004; Snethlage 2005; Pinto Guerra 2008). Some recent translations have been useful (Caneva, Nugari, and Salvadori 2008), but they are relatively rare. Standards committees (ISO, CEN, ASTM, RILEM) have also brought a needed level of integration, especially at the European level, for example, Technical Committee 346 (Fassina 2008).

The Internet has made a huge impact. Enormous amounts of information on stone conservation are more readily available and accessible than ever before. Nonetheless, many of the tools needed to access that information are still lacking, such as a citation index, more-comprehensive databases of conservation-related research material, and wider electronic distribution of conference proceedings, past and present. Opportunities, such as a Wiki, for more widespread conservation community feedback and contributions to conservation research, would also be valuable. Without these tools, it is a time-consuming challenge to find high-quality research that advances the field.[3] Addressing these gaps would accelerate the return on our investment in research and would add new tools to the stone conservator's kit.

So far, so good—there have been many changes for the better. Nonetheless, many of the issues that dogged research in 1994 (when the first edition of this volume was written) are still with us: the tendency to publish research in conference proceedings that are not refereed and not widely available, the variable quality of research, the multiplicity of conferences, and the ongoing reinvention of the wheel due to the difficulties of accessing previous work in the field. National funding cycles for stone conservation still tend toward large-scale interventions on sites "in crisis," while funding for routine maintenance remains in short supply and funding for long-term research is even more difficult to come by.

Long-tem funding is of particular importance, given the need to evaluate and document treatments over long periods of time—much longer than the duration of individual research projects. The nonuniversity institutions have an important role here, facilitating long-term applied research. Discrete or isolated measurements and projects are of limited utility. A more nuanced understanding of material behavior and the effects of conservation interventions over time is essential for balanced and effective decision making, and this can be achieved only in the context of long-term research.

What is the relevance of all this for the stone conservator? It sometimes seems that the tool kit of a stone conservator has not changed much in the past two decades and may even contain fewer options now than then due to regulations, environmental concerns, health concerns, compatibility concerns, and lessons learned from unintended consequences. As conservators begin to specialize more in recording, investigation, and characterization and less in treatments, and stone replacement becomes more common, this raises the question: are treatments still needed? A colleague answered this question by suggesting, "the field has changed, but the stone has not, and in many cases it is crumbling."

So yes, there is still a role for both preventive and active interventions in the conservator's tool kit, and important new treatment options have been developed over the past fifteen years. Examples include: 1) more-advanced methods of controlling clay swelling of stone, 2) coupling agents for limestone consolidation, 3) latex solutions and laser systems for stone cleaning, 4) improved poulticing methods, 5) water-based hydrophobic coatings, 6) less-brittle silane consolidants, and 7) nanoparticle solutions of lime for consolidation of fragile stone surfaces (Table 7.1). Inevitably, there is a time lag between development and widespread application, and there is an onus on all researchers—whether sci-

Table 7.1
Stone Conservator's Tool Kit

Interventions with wide application	• Nano-lime particles suspended in alcohol • Water-based hydrophobic coatings • Spray-on latex for cleaning architectural interiors • Portable, large-scale laser systems for cleaning • Bioremediation
Interventions that are under development	• Coupling agents for limestone consolidation • Improved poulticing methods • Treatments for clay swelling of stone • Nano-particle-modified silane consolidants; calcium alkoxides; calcium phosphate or oxalate treatments • Nanotechnology cleaning agents
Preventive conservation	• Microclimate stabilization and shelters • Mitigation of rapid environmental fluctuations for immovable cultural property • Environmental control for salt-laden structures based on computer models and observations • Wind fences, trees, reburial, etc.
Documentation tools	• 3D laser scanning to quantify surfaces • Quantitative calibration of digital color images • Solving the lighting problem—repeat photography – PTM images – Color matching

entists or conservators—to ensure that their findings are implemented appropriately. But the tool kit has undoubtedly changed, and further changes are on their way, as the tool kit for researchers (Table 7.2) has added new tools, including Focused Ion Beam/Environmental Scanning Electron Microscopy (FIB/ESEM), cryo-scanning electron microscopy (cryo-SEM), and wet-scanning transmission electron microscopy (wet-STEM), among many others. The context in which these conservation tool kits are used now includes the impacts of climate change, the Internet, and the rapid development of nanotechnology (Table 7.3).

Beyond the basic tool kit, there are signs of a new maturity in the field of stone conservation. An awareness of the unintended consequences of some earlier interventions has, for example, resulted in a more cautious and incremental approach in the current generation of stone conservators.

Table 7.2
Conservation Scientist's Took Kit

Tools for damage monitoring	• Laser interferometry • Laser scanning • Real-time crack monitoring • Time-lapse systems • Linear Variable Differential Transformer (LVDT)
Research tools	• NMR • FIB/ESEM, cryo-SEM, wet-STEM • CT-scanning • Thermal analysis • Damage models • Heat and moisture transport models

Table 7.3
Current Trends in Conservation Research

- Impacts of climate change
- Rare events versus routine damage
- Use of volunteers in conservation assessments
- Internet (access to research and commentary)
- Biomimetic surfaces
- Nanotechnology

In many parts of the world we are now less likely to see the heavy-handed use of biocides, waterproofing agents, and consolidants, and more likely to see emphasis on careful documentation, monitoring, regular maintenance, control of moisture, selective use of waterproofing agents and consolidants, stone replacement, and the design of minimally invasive treatments. The tradition of regular attention and maintenance is finally being seen as a more realistic alternative to the dream of a cure-all, silver-bullet stone preservative. The preservation of our heritage in stone will ultimately benefit from our growing understanding of material behavior (Torraca 2009) and the maintenance necessary to sustain long-term performance (Brand 1995).

CONCLUSION

This volume opened with the suggestion that our knowledge of stone was outstripping the practical application of that knowledge to stone conservation problems. We have seen that there have, indeed, been major advances in our knowledge of stone behavior. This strong scientific foundation has also been accompanied by an encouraging number of new conservation treatments, methods, and tools.

The key challenge for the future is that resources for applied research, technology transfer, and long-term testing are needed. While progress in these areas has undoubtedly been evident over the last fifteen years, structural gaps remain between researchers and practitioners and between the old assumptions and rapidly evolving new knowledge. Scarce resources for stone research are not always being applied to best use. This may be a useful moment to rethink the structural problems inherent in traditional approaches to conservation projects and funding. In order to preserve our heritage in stone, it is time to build support for larger-scale and longer-term research and technology transfer projects. In a number of cases, we have exciting solutions to stone conservation problems, but we do not have the resources to properly test and implement these solutions.

Notes

1 EH = English Heritage; NPS = National Park Service, USA; TNO = Netherlands Technical Organization.

2 Globish is a term proposed by a French academic for a subset of 1,500 English words often used for global communication (see http://en.wikipedia.org/wiki/Globish). The term "globish" is a blend of "global" and "English."

3 Some more recent references in this book's bibliography contain digital object identifiers (DOIs), which can be used to link to online resources using the following Web site: http://dx.doi.org/.

References

Åberg, G., H. Stray, and E. Dahlin. 1999. Impact of pollution at a Stone Age rock art site in Oslo, Norway, studied using lead and strontium isotopes. *Journal of Archaeological Science* 26 (12): 1483–88.

Abu-Jaber, N., Z. Al Saad, and M. Al Qudah. 2006. Landscape, geology and lithology of the Northern Al Jafr basin, Jordan. In *Conservation of Ancient Stone Quarry Landscapes in the Eastern Mediterranean: First QuarryScapes Symposium, October 15–17, 2006, Antalya, Turkey: Extended Abstract Collection,* ed. P. Degryse, 5–6. Trondheim, Norway: QuarryScapes Project. http://www.quarryscapes.no/text/Workshops/Proceedings_1stQS_symposium.pdf#page=5.

Adamo, P., and P. Violante. 2000. Weathering of rocks and neogenesis of minerals associated with lichen activity. *Applied Clay Science* 16 (5–6): 229–56.

Adams, A. E., and W. S. MacKenzie. 1998. *A Colour Atlas of Carbonate Sediments and Rocks Under the Microscope.* London: Manson.

Adolfs, N. C. 2007. Die Anwendung von Calciumhydroxid-Sol als Festigungsmittel für historische Putze: Erste Versuche und deren Überprüfung. Diplomarbeit, Fachhochschule, Cologne, Germany. http://www.icvbc.cnr.it/drilling/publications/tesi%20in%20tedesco%20DRMS.pdf.

Agarossi, G., R. Ferrari, and M. Monte. 1990. Biocides in the control of biodeterioration. In *The Conservation of Monuments in the Mediterranean Basin: The Influence of Coastal Environment and Salt Spray on Limestone Marble: Proceedings of the 1st International Symposium, Bari, 7–10 June 1989 = La conservazione dei monumenti nel bacino del Mediterraneo: Influenza dell'ambiente costiero e dello spray marino sulla pietra calcarea e sul marmo: Atti del 1° simposio internazionale, Bari, 7–10 giugno 1989,* ed. F. Zezza, 511–17. Brescia, Italy: Grafo Edizioni.

Aglietto, M., E. Passaglia, E. Taburoni, F. Ciardelli, C. Botteghi, U. Matteoli, S. Paganelli, R. Arbizzani, and V. Fassina. 1993. A new class of fluorinated acrylic polymers: Protective materials for stone. In *10th Triennial Meeting, ICOM Committee for Conservation, Washington, DC, 22–27 August 1993: Preprints,* ed. J. Bridgland, 553–58. Paris: ICOM Committee for Conservation; Lawrence, KS: Allen Press (distributor).

Agnew, N., and S. Maekawa. 1999. Preserving Nefertari's legacy. *Scientific American* 281 (4): 52–57.

Agnew, N., S. Maekawa, R. Coffman, and J. Meyer. 1996. Evaluation of the performance of a lightweight modular site shelter: Quantitative meteorological data and protective indices for the "hexashelter." *Conservation and Management of Archaeological Sites* 1 (3): 139–50.

Ahmon, J. 2004. The application of short-range 3D laser scanning for archaeological replica production: The Egyptian tomb of Set 1. *Photogrammetric Record* 19 (106): 111–27.

Aira, N., V. Jurado, B. Silva, and B. Prieto. 2007. Gas chromatography applied to cultural heritage: Analysis of dark patinas on granite surfaces. *Journal of Chromatography A* 1147 (1): 79–84.

Aires-Barros, L., A. Maurício, and C. Figueiredo. 1994. Profilometry and image analysis applications to "in situ" study of monuments stone decay phenomena. In *Conservation of Monuments in the Mediterranean Basin: Stone Monuments,*

Methodologies for the Analysis of Weathering and Conservation; Proceedings of the 3rd International Symposium, Venice, 22–25 June 1994 = La conservazione dei monumenti nel bacino del Mediterraneo: Materiali lapidei e monumenti, metodologie per l'analisi del degrado e la conservazione; Atti del 3° simposio internazionale, Venezia, 22–25 giugno 1994, ed. V. Fassina, H. Ott, and F. Zezza, 19–24. Venice: Soprintendenza ai beni artistici e storici di Venezia.

Albero, S., C. Giavarini, M. L. Santarelli, and A. Vodret. 2004. CFD modeling for the conservation of the Gilded Vault Hall in the Domus Aurea. *Journal of Cultural Heritage* 5 (2): 197–203.

Albertsen, J. S., ed. 2008. *Salt Weathering on Buildings and Stone Sculptures, 22–24 October 2008, The National Museum, Copenhagen, Denmark* [Proceedings from the international conference]. Lyngby: Technical University of Denmark, Department of Civil Engineering. http://www.swbss.dk/.

Alessandrini, G., and M. Laurenzi Tabasso. 1999. Conservation of cultural property in Italy: The UNI-NORMAL Committee for the Definition of Technical Standards. In *The Use of and Need for Preservation Standards in Architectural Conservation,* ed. L. B. Sickles-Taves, 24–30. ASTM Special Technical Publication 1355. W. Conshohocken, PA: ASTM.

Alessandrini, G., and A. Pasetti. 2004. L'elenco ragionato delle norme UNI-NorMaL [Annotated list of UNI-NorMaL norms]. *Arkos: scienza e restauro dell'architettura* 5 (6): 66–70.

Alessandrini, G., L. Toniolo, A. Antonioli, A. Di Silvestro, F. Piacenti, S. Righini Ponticelli, and L. Formica. 1993. On the cleaning of deteriorated stone minerals. In *Conservation of Stone and Other Materials: Proceedings of the International RILEM/UNESCO Congress "Conservation of Stone and Other Materials: Research – Industry – Media," UNESCO Headquarters, Paris, June 29–July 1, 1993,* ed. M.-J. Thiel, 503–11. RILEM Proceedings 21. London and New York: E & FN Spon.

Alfano, G., C. Chiancarella, E. Cirillo, I. Fato, and F. Martellotta. 2006. Long-term performance of chemical damp-proof courses: Twelve years of laboratory testing. *Building and Environment* 41 (8): 1060–69.

Allanbrook, T., and K. C. Normandin. 2007. The restoration of the Fifth Avenue facades of the Metropolitan Museum of Art. *APT Bulletin* 38 (4): 45–53.

Allemand, L., and P. G. Bahn. 2005. Best way to protect rock art is to leave it alone. *Nature* 433 (7028): 800.

Allen, G. 2007. Building rescue. *Materials World* 15 (6): 27–28.

Alonso, F. J., R. M. Esbert, J. Alonso, and J. Ordaz. 1994. Saline spray action on a treated dolomitic stone. In *Conservation of Monuments in the Mediterranean Basin: Stone Monuments, Methodologies for the Analysis of Weathering and Conservation; Proceedings of the 3rd International Symposium, Venice, 22–25 June 1994 = La conservazione dei monumenti nel bacino del Mediterraneo: Materiali lapidei e monumenti, metodologie per l'analisi del degrado e la conservazione; Atti del 3° simposio internazionale, Venezia, 22–25 giugno 1994,* ed. V. Fassina, H. Ott, and F. Zezza, 867–70. Venice: Soprintendenza ai beni artistici e storici di Venezia.

Ambrosi, M., P. Baglioni, P. R. David, L. Dei, R. Giorgi, C. Lalli, G. Lanterna, A. Mairani, M. Matteini, M. Rizzi, and G. Schonhaut. 2000. Inorganic consolidants and protectives for architectonic surfaces: Experimental tests on Santa Prisca Church apse in Rome. In *2nd International Congress on Science and Technology for the Safeguard of Cultural Heritage in the Mediterranean Basin, 5–9 July 1999, Paris, France: Proceedings,* ed. A. Guarino, vol. 2, 873–77. Paris: Éd. scientifiques et médicales Elsevier.

Ambrosi, M., L. Dei, R. Giorgi, C. Neto, and P. Baglioni. 2001. Colloidal particles of Ca(OH)2: Properties and applications to restoration of frescoes. *Langmuir* 17 (14): 4251–55.

Anagnostidis, K., C. K. Gehrmann, M. Gross, W. E. Krumbein, S. Lisi, A. Pantazidou, C. E. Urzi, and M. Zagari. 1992. Biodeterioration of marbles of the Parthenon and Propylaea, Acropolis, Athens: Associated organisms, decay and treatment suggestions. In *La conservation des monuments dans le bassin méditerranéen: Actes du 2ème Symposium international = The Conservation of Monuments in the Mediterranean Basin: Proceedings of the 2nd International Symposium,* ed. D. Decrouez, J. Chamay, and F. Zezza, 305–25. Geneva: Ville de Genève, Muséum d'histoire naturelle, and Musée d'art et d'histoire.

Andersson, T. 1986. Preservation and restoration of rock carvings and rune-stones. In *Case Studies in the Conservation of Stone and Wall Paintings: Preprints of the Contributions to the Bologna Congress, 21–26 September 1986,* ed. N. S. Brommelle and P. Smith, 133–37. London: International Institute for Conservation of Historic and Artistic Works.

André, M. F., S. Etienne, D. Mercier, F. Vautier, and O. Voldoire. 2008. Assessment of sandstone deterioration at Ta Keo temple (Angkor): First results and future prospects. *Environmental Geology* 56 (3–4): 677–88.

Andreotti, A., I. Bonaduce, M. Perla Colombini, F. Modugno, and E. Ribechini. 2009. A diagnosis of the yellowing of the marble high reliefs and the black decorations in the chapel of the tomb of Saint Anthony (Padua, Italy). In Art and Cultural Heritage. Special issue, *International Journal of Mass Spectrometry* 284 (1–3): 123–30.

Andrew, C. M., M. Young, and K. Tonge. 1994. *Stone Cleaning: A Guide for Practitioners.* Aberdeen: Robert Gordon University; Edinburgh: Historic Scotland.

Anselmi, C., B. Doherty, and F. Presciutti. 2008. Evaluation of cyclododecane as a temporary consolidant for cultural heritage materials. In *Stone Consolidation in Cultural Heritage: Research and Practice; Proceedings of the International Symposium, Lisbon, 6–7 May 2008,* ed. J. Delgado Rodrigues and J. M. Mimoso, 203–12. Lisbon: LNEC (Laboratório Nacional de Engenharia Civil).

Antonelli, V. 1979. Il restauro della Porta della Carta in Venezia. In *Deterioramento e conservazione della pietra: Atti del 3° Congresso internazionale, Venezia, 24–27 ottobre 1979, Fondazione "Giorgio Cini," Isola di S. Giorgio Maggiore = Deterioration and Preservation of Stones: Proceedings of the 3rd International Congress, Venice, 24–27 October 1979 = La déterioration et la préservation de la pierre: Actes du 3ème Congrès international, Venise 24–27 octobre 1979,* 629–44. Padua: Università degli Studi di Padova, Istituto di Chimica Industriale.

Antonioli, P., G. Zapparoli, P. Abbruscato, C. Sorlini, G. Ranalli, and P. G. Righetti. 2005. Art-loving bugs: The resurrection of Spinello Aretino from Pisa's cemetery. *Proteomics* 5 (9): 2453–59.

Arnold, A. 1996. Altération et conservation d'oeuvres culturelles en matériaux poreux affectés par des sels. In *Le dessalement des matériaux poreux: 7es Journées d'études de la SFIIC, Poitiers, 9–10 mai 1996,* 3–20. Champs-sur-Marne: SFIIC.

Arnold, A., and K. Zehnder. 1991. Monitoring wall paintings affected by soluble salts. In *The Conservation of Wall Paintings: Proceedings of a Symposium Organized by the Courtauld Institute of Art and the Getty Conservation Institute, London, July 13–16, 1987,* ed. S. Cather, 103–35. Marina del Rey, CA: Getty Conservation Institute. http://www.getty.edu/conservation/publications/pdf_publications/wall_paintings.pdf.

Arocena, J. M., K. Hall, and I. Meiklejohn. 2008. Minerals provide tints and possible binder/extender in pigments in San rock paintings. *Geoarchaeology* 23 (2): 293–304.

Ascaso, C., J. Wierzchos, V. Souza-Egipsy, A. de Los Ríosa, and J. Delgado Rodrigues. 2002. In situ evaluation of the biodeteriorating action of microorganisms and the effects of biocides on carbonate rock of the Jeronimos Monastery (Lisbon). *International Biodeterioration and Biodegradation* 49 (1): 1–12.

Ashurst, J. 1998. The cleaning and treatment of limestone by the lime method. In *Conservation of Building and Decorative Stone,* ed. J. Ashurst and F. G. Dimes, vol. 2, 169–76. Butterworth-Heinemann Series in Conservation and Museology. London and Boston: Butterworth-Heinemann.

Ashurst, J. 2007. *Conservation of Ruins.* Butterworth-Heinemann Series in Conservation and Museology. London and Burlington, MA: Butterworth-Heinemann.

Ashurst, J., and N. Ashurst. 1988. *Practical Building Conservation.* Vol. 1, *Stone Masonry.* English Heritage Technical Handbook. New York: Halsted Press.

Ashurst, J., and F. G. Dimes, eds. 1998. *Conservation of Building and Decorative Stone.* Butterworth-Heinemann Series in Conservation and Museology. Oxford and Woburn, MA: Butterworth-Heinemann.

Ashurst, N. 1994. *Cleaning Historic Buildings*. London: Donhead Publishing.

Aslan, Z. M. 2007. The design of protective structures for the conservation and presentation of archaeological sites. PhD diss., University College, London.

Asmus, J. F., G. Guattari, L. Lazzarini, G. Musumeci, and R. F. Wuerker. 1973. Holography in the conservation of statuary. *Studies in Conservation* 18 (2): 49–63.

Associated Press (AP). 2009. Saving the art of Lascaux. *Washington Post,* February 27, 2009, A 10. http://www.washingtonpost.com/wp-dyn/content/article/2009/02/26/AR2009022603630_pf.html.

Atlas, R. M., A. N. Chowdhury, and K. L. Gauri. 1988. Microbial calcification of gypsum-rock and sulfated marble. *Studies in Conservation* 33 (3): 149–53.

Atzeni, C., U. Sanna, and N. Spanu. 2006. Some mechanisms of microstructure weakening in high-porous calcareous stones. *Materials and Structures/Materiaux et constructions* 39 (5): 525–31.

Auras, M. 1993. Stone conservation by impregnation with polyurethane. In *Conservation of Stone and Other Materials: Proceedings of the International RILEM/UNESCO Congress "Conservation of Stone and Other Materials: Research—Industry—Media," UNESCO Headquarters, Paris, June 29–July 1, 1993,* ed. M.-J. Thiel, 717–24. RILEM Proceedings 21. London and New York: E & FN Spon.

Auras, M. 2008. Poultices and mortars for salt contaminated masonry and stone objects. In *Salt Weathering on Buildings and Stone Sculptures: 22–24 October 2008, The National Museum Copenhagen, Denmark: [Proceedings From the International Conference]*, ed. Judith Selk Albertsen, 197–217. Lyngby: Technical University of Denmark, Department of Civil Engineering.

Ausset, P., R. Lefèvre, J. Philippon, and C. Venet. 1992. Large-scale distribution of fly-ash particles inside weathering crusts on calcium carbonate substrates: Some examples on French monuments. In *La conservation des monuments dans le bassin méditerranéen: Actes du 2ème symposium international = The Conservation of Monuments in the Mediterranean Basin: Proceedings of the 2nd International Symposium,* ed. D. Decrouez, J. Chamay, and F. Zezza, 121–39. Geneva: Ville de Genève, Muséum d'histoire naturelle, and Musée d'art et d'histoire.

Baedecker, P. A., and M. M. Reddy. 1993. The erosion of carbonate stone by acid rain: Laboratory and field investigations. *Journal of Chemical Education* 70 (2): 104–8.

Baer, N. S. 1991. Assessment and management of risks to cultural property. In *Science, Technology, and European Cultural Heritage: Proceedings of the European Symposium, Bologna, Italy, 13–16 June 1989,* ed. N. S. Baer, C. Sabbioni, and A. I. Sors, 27–36. Oxford and Boston: Published for the Commission of the European Communities by Butterworth-Heinemann Publishers.

Baer, N. S., and R. Snethlage, eds. 1997. *Saving Our Architectural Heritage: The Conservation of Historic Stone Structures; Report of the Dahlem Workshop on Saving Our Architectural Heritage, The Conservation of Historic Stone Structures, Berlin, March 3–8, 1996*. Dahlem Workshop Reports. Chichester and New York: John Wiley & Sons.

Baer, N. S., and F. Snickars, eds. 2001. *Rational Decision-Making in the Preservation of Cultural Property*. Dahlem Workshop Reports 86. Berlin: Dahlem University Press.

Bahn, P. 2008. Killing Lascaux: Inept bureaucrats and creeping fungi are destroying the world's most famous cave paintings. *Archaeology* 61 (3): 18.

Bai, Y., G. E. Thompson, and S. Martinez-Ramirez. 2006. Effects of NO_2 on oxidation mechanisms of atmospheric pollutant SO_2 over Baumberger sandstone. *Building and Environment* 41 (4): 486–91.

Bakkevig, S. 2004. Rock art preservation: Improved and ecology-based methods can give weathered sites prolonged life. *Norwegian Archaeological Review* 37 (2): 65–81.

Bala'awi, F. 2008. Wind speed and salt simulation tests: Toward a more comprehensive approach. In *Salt Weathering on Buildings and Stone Sculptures, 22–24 October 2008, The National Museum, Copenhagen, Denmark* [Proceedings from the international conference], ed. J. S. Albertsen, 41–50. Lyngby: Technical University of Denmark, Department of Civil Engineering. http://www.swbss.dk/.

Balboni, E., R. M. Espinosa-Marzal, E. Doehne, and G. W. Scherer. Forthcoming. Can drying and re-wetting of magnesium sulphate salts lead to damage of stone? *Environmental Earth Sciences*.

Ball, J., M. E. Young, and R. A. Laing. 2000. A simple technique for rapid field assessment of stone decay on buildings. In *Proceedings of the 9th International Congress on Deterioration and Conservation of Stone, Venice, June 19–24, 2000,* ed. V. Fassina, vol. 2, 13–21. Amsterdam and New York: Elsevier.

Barajas, M., E. Lima, V. H. Lara, J. V. Negrete, C. Barragáín, C. Malváez, and P. Bosch. 2009. Effect of organic and inorganic consolidation agents on Tlaltecuhtli monolith. *Journal of Archaeological Science* 36 (10): 2244–52.

Barff, F. S. (inventor). 1860. Stone, Artificial Stone, Preserving Stone, Colouring. British Patent 2608, issued 1860.

Barnett, T., A. Chalmers, M. Díaz-Andreu, P. Longhurst, G. Ellis, K. Sharpe, and I. Trinks. 2005. Scan 3D au laser pour enregistrer et surveiller l'érosion de l'art rupestre = 3D laser scanning for recording and monitoring rock art erosion. *INORA: International Newsletter on Rock Art* (41): 25–28. http://www.bradshawfoundation.com/inora/.

Barnett, T., and M. Díaz-Andreu. 2005. Knowledge capture and transfer in rock art studies: Results of a questionnaire on rock art decay in Britain. *Conservation and Management of Archaeological Sites* 7 (1): 35–48.

Bassier, C. 1989. Le rayonnement ultra-violet dans la désinfection des matériaux de nature ou d'origine minérale: Exemples d'application = Ultraviolet radiation in the disinfection of materials of a mineral nature or origin: Examples of its application. In *Patrimoine culturel et altérations biologiques: Actes des journées d'étude de la S.F.I.I.C, Poitiers, 17 et 18 novembre 1988,* 201–6. Champs-sur-Marne: Section française de l'Institut international de conservation.

Bastian, F., and C. Alabouvette. 2009. Lights and shadows on the conservation of a rock art cave: The case of Lascaux Cave. *International Journal of Speleology* 38 (1): 55–60.

Bastian, F., C. Alabouvette, and C. Saiz-Jimenez. 2009. Bacteria and free-living amoeba in the Lascaux Cave. *Research in Microbiology* 160 (1): 38–40.

Bastian, F., C. Alabouvette, V. Jurado, and C. Saiz-Jimenez. 2009. Impact of biocide treatments on the bacterial communities of the Lascaux Cave. *Naturwissenschaften* 96 (7): 863–68.

Bastian, F., G. Orial, A. François, and C. Alabouvette. 2007. La grotte de Lascaux: Un écosystème complexe où bactéries et champignons interagissent. *Biofutur* 26 (283): 28–31.

Batarda-Fernandes, A., and J. Delgado Rodrigues. 2008. Stone consolidation experiments in rock art outcrops at the Côa Valley Archaeological Park, Portugal. In *Stone Consolidation in Cultural Heritage: Research and Practice; Proceedings of the International Symposium, Lisbon, 6–7 May 2008,* ed. J. Delgado Rodrigues and J. M. Mimoso, 111–20. Lisbon: LNEC (Laboratório Nacional de Engenharia Civil).

Bates, K. T., F. Parity, P. L. Manning, D. Hodgetts, B. Vila, O. Oms, À. Galobart, and R. L. Gawthorpe. 2008. High-resolution LiDAR and photogrammetric survey of the Fumanya dinosaur tracksites (Catalonia): Implications for the conservation and interpretation of geological heritage sites. *Journal of the Geological Society* 165 (1): 115–27.

Beaubien, H. F., T. E. Collas, C. E. Magee, S. B. Peschken, E. F. Rosenthal, M. E. Svoboda, and C. Mei-An Tsu. 1999. Desalination parameters for Harappan ceramics, Part 2. In *Objects Specialty Group Postprints: Proceedings of the Objects Specialty Group Session, June 11, 1999, 27th Annual Meeting, American Institute for Conservation of Historic and Artistic Works, St. Louis,* ed. V. Greene and E. Kaplan, vol. 6, 77–93. Washington, DC: American Institute for Conservation of Historic and Artistic Works, Objects Specialty Group.

Bell, J., J.-P. Jerôme, P. Sawyer, V. Magar, and N. Stanley-Price. 1996. Stereophotogrammetric recording of rock art at the Cueva de El Ratón, Baja California, Mexico. In *11th Triennial Meeting, ICOM Committee for Conservation, Edinburgh, 1–6 September 1996: Preprints,* ed. J. Bridgland, 454–57. London: James & James (Science Publishers) Ltd.

Benarie, M. 1991. The establishment and use of damage functions. In *Science, Technology, and European Cultural Heritage: Proceedings of the European*

Symposium, Bologna, Italy, 13–16 June 1989, ed. N. S. Baer, C. Sabbioni, and A. I. Sors, 214–20. Oxford and Boston: Published for the Commission of the European Communities by Butterworth-Heinemann Publishers.

Bernabe, E., P. Bromblet, and M. Robert. 1995. Role de la cristallisation du natron dans la desagregation sableuse d'un monument granitique en Bretagane = Role of natron crystallization on granular disintegration of a granitic monument in Brittany. *Comptes Rendus Academie des Sciences: Serie IIA, Sciences de la Terre et des Planetes* 320 (7): 571–78.

Bescher, E., and J. D. Mackenzie. 2003. Sol-gel coatings for the protection of brass and bronze. *Journal of Sol-Gel Science and Technology* 26 (1–3): 1223–26.

Binda, L., ed. 2007. *Learning from Failure: Long-Term Behaviour of Heavy Masonry Structures.* International Series on Advances in Architecture. Southampton: WIT.

Binda, L., M. Lualdi, and A. Saisi. 2007. Non-destructive testing techniques applied for diagnostic investigation: Syracuse Cathedral in Sicily, Italy. *International Journal of Architectural Heritage* 1 (4): 380–402.

Binda, L., A. Saisi, C. Tiraboschi, S. Valle, C. Colla, and M. Forde. 2003. Application of sonic and radar tests on the piers and walls of the Cathedral of Noto. *Construction and Building Materials* 17 (8): 613–27.

Bionda, D. 2004. Methodology for the preventive conservation of sensitive monuments: Microclimate and salt activity in a church. In *Proceedings of the 10th International Congress on Deterioration and Conservation of Stone, Stockholm, June 27–July 2, 2004,* ed. Daniel Kwiatkowski and Runo Löfvendahl, vol. 2, 627–34. Stockholm: ICOMOS, Sweden.

Biscontin, G., P. Maravelaki, E. Zendri, A. Glisenti, and E. Tondello. 1993. Investigation into the interaction between aqueous and organic solvent protection and building materials. In *Conservation of Stone and Other Materials: Proceedings of the International RILEM/UNESCO Congress "Conservation of Stone and Other Materials: Research – Industry – Media," UNESCO Headquarters, Paris, June 29–July 1, 1993,* ed. M.-J. Thiel, 689–96. RILEM Proceedings 21. London and New York: E & FN Spon.

Bjelland, T., and S. Ekman. 2005. Fungal diversity in rock beneath a crustose lichen as revealed by molecular markers. *Microbial Ecology* 49 (4): 598–603.

Bjelland, T., and I. H. Thorseth. 2002. Comparative studies of the lichen-rock interface of four lichens in Vingen, western Norway. *Chemical Geology* 192 (1–2): 81–98.

Bjelland, T., K. Sognes, J. Goldhahn, and S. Bakkevig. 2005. Comments on Sverre Bakkevig, Rock art preservation: Improved and ecology-based methods can give weathered sites prolonged life. *Norwegian Archaeological Review* 38 (1): 49–67.

Black, S. N., L. A. Bromley, D. Cottier, R. J. Davey, B. Dobbs, and J. E. Rout. 1991. Interactions at the organic/inorganic interface: Binding motifs for phosphonates at the surface of barite crystals. *Journal of the Chemical Society, Faraday Transactions* 87 (20): 3409–14.

Blanc, A., and C. Lorenz. 1992. Les marbres et les roches ornementales dans les monuments. In *La conservation de la pierre monumentale en France,* ed. J. Philippon, D. Jeannette, and R.-A. Lefèvre, vol. 2, *Conservation du patrimoine,* 23–26. Paris: Presses du CNRS (Ministère de la culture et de la communication).

Blanc, A., and C. Lorenz. 1988. Etude geologique des anciennes carrieres de Paris: Son utilite pour la connaissance et la restauration des monuments. In *The Engineering Geology of Ancient Works, Monuments and Historical Sites: Preservation and Protection; Proceedings of an International Symposium Organized by the Greek National Group of IAEG, Athens, 19–23 September 1988,* ed. P. G. Marinos and G. Koukis, 639–47. Rotterdam and Brookfield, VT: Balkema.

Bläuer Böhm, C. 2004. Techniques and tools for conservation investigations. In *Proceedings of the 10th International Congress on Deterioration and Conservation of Stone, Stockholm, June 27–July 2, 2004,* ed. D. Kwiatkowski and R. Löfvendahl, vol. 2, 549–59. Stockholm: ICOMOS Sweden.

Bläuer Böhm, C. 2005. Quantitative salt analysis in conservation of buildings. *Restoration of Buildings and Monuments: An International Journal = Bauinstandsetzen und Baudenkmalpflege: Eine internationale Zeitschrift* 11 (6): 409–18.

Bläuer, C., and A. Kueng. 2007. Examples of microscopic analysis of historic mortars by means of polarising light microscopy of dispersions and thin sections. Special issue, *Materials Characterization* 58 (11–12): 1199–207.

Bloxam, E. 2006. From complex data to simple transmission: Modelling the significance of ancient quarry landscapes. In *Conservation of Ancient Stone Quarry Landscapes in the Eastern Mediterranean: First QuarryScapes Symposium, October 15–17, 2006, Antalya, Turkey: Extended Abstract Collection,* ed. P. Degryse, 27–29. Trondheim, Norway: QuarryScapes Project. http://www.quarryscapes.no/text/Workshops/Proceedings_1stQS_symposium.pdf#page=27.

Boos, M., J. Grobe, G. Hilbert, and J. Müller-Rochholz. 1996. Modified elastic silicic-acid ester applied on natural stone and tests of their efficiency. In *Proceedings of the 8th International Congress on Deterioration and Conservation of Stone: Berlin, 30. Sept.–4. Oct. 1996,* ed. J. Riederer, 1179–85. Berlin: Möller Druck und Verlag.

Borrelli, E., and A. Urland. 1999. *ARC Laboratory Handbook*. Rome: ICCROM. http://www.iccrom.org/pdf/ICCROM_14_ARCLabHandbook00_en.pdf.

Bourgès, A. 2006. Holistic correlation of physical and mechanical properties of selected natural stones for assessing durability and weathering in the natural environment. PhD diss., Universität München. http://edoc.ub.uni-muenchen.de/5505/1/Bourges_Ann.pdf.

Bourgès, A., and V. Vergès-Belmin. 2008a. A new methodology to determine rheologic behavior and mechanical properties of desalination poultices. In *Proceedings of the 11th International Congress on Deterioration and Conservation of Stone, 15–20 September 2008, Toruń, Poland,* ed. J. W. Lukaszewicz and P. Niemcewicz, 581–88. Toruń, Poland: Nicolaus Copernicus University.

Bourgès, A., and V. Vergès-Belmin. 2008b. Comparison and optimization of five desalination systems on the inner walls of Saint Philibert Church in Dijon, France. In *Salt Weathering on Buildings and Stone Sculptures, 22–24 October 2008, The National Museum Copenhagen, Denmark* [Proceedings from the international conference], ed. J. S. Albertsen, 29–40. Lyngby: Technical University of Denmark, Department of Civil Engineering. http://www.design.upenn.edu/files/14-Bourges__Verges_Belmin_Desalination_SWBSS_2008.pdf.

Bourgès, A., K. T. Fehr, S. Simon, and R. Snethlage. 2008. Correlation between the micro-structure and the macroscopic behavior of sandstones. *Restoration of Buildings and Monuments: An International Journal = Bauinstandsetzen und Baudenkmalpflege: Eine internationale Zeitschrift* 14 (3): 157–66.

Bourguignon, E., F. Bertrand, C. Moreau, P. Coussot, and N. Shahidzadeh-Bonn. 2008. Desalination of model stones by poulticing. In *Proceedings of the 11th International Congress on Deterioration and Conservation of Stone, 15–20 September 2008, Toruń, Poland,* ed. J. W. Lukaszewicz and P. Niemcewicz, 803–10. Toruń, Poland: Nicolaus Copernicus University.

Boutin, F. 2001. Comparative study of the efficiency of protective treatments applied to stone. In *Surface Technology with Water Repellent Agents: Proceedings of Hydrophobe III, Third International Conference on Surface Technology with Water Repellent Agents, Universität Hannover, Germany, September 25 and 26, 2001,* ed. K. Littman and A. E. Charola, 233–44. Freiburg: Aedificatio Verlag.

Bowley, M. J. 1975. *Desalination of Stone: A Case Study*. Building Research Establishment Current Paper 46/75. [Watford, England]: Building Research Establishment.

Bracci, S., J. Delgado Rodrigues, A. Ferreira Pinto, M. Matteini, D. Pinna, S. Porcinai, B. Sacchi, and B. Salvadori. 2008. Development and evaluation of new treatments for the conservation of outdoor stone monuments. In *Proceedings of the 11th International Congress on Deterioration and Conservation of Stone, 15–20 September 2008, Toruń, Poland,* ed. J. W. Lukaszewicz and P. Niemcewicz, 811–18. Toruń, Poland: Nicolaus Copernicus University.

Bracci, S., and M. J. Melo. 2003. Correlating natural ageing and Xenon irradiation of Paraloid® B72 applied on stone. *Polymer Degradation and Stability* 80 (3): 533–41.

Brajer, I., and N. Kalsbeek. 1999. Limewater absorption and calcite crystal formation on a limewater-impregnated secco wall painting. *Studies in Conservation* 44 (3): 145–56.

Brand, S. 1995. *How Buildings Learn: What Happens After They're Built*. New York: Penguin Books.

Brimblecombe, P. 1992. History of air pollution and deterioration of heritage. In *Weathering and Air Pollution: Primo corso della Scuola universitaria C.U.M. conservazione dei monumenti, Lago di Garda (Portese), Venezia, Milano, 2–9 settembre 1991*, ed. F. Zezza, 23–32. Bari: Mario Adda Editore.

Brimblecombe, P. 2001. Organics and metals in urban aerosols: Research directions. *Journal of Aerosol Science* 32 (suppl. 1): 319–20.

Brimblecombe, P., and C. M. Grossi. 2007. Damage to buildings from future climate and pollution. *APT Bulletin* 38 (2–3): 13–18.

British Standards Institution (BSI). 2000. *BS 8221-2:2000 Code of Practice for Cleaning and Surface Repair of Buildings: Surface Repair of Natural Stones, Brick and Terracotta*. London: BSI.

Brokerhof, A., V. Meul, S. Michalski, and J. L. Pedersoli. 2007. Advancing research in risk management applications to cultural property. *ICCROM Newsletter* (33): 10–11. http://www.iccrom.org/pdf/ICCROM_newsl33-2007_en.pdf.

Bromblet, P., and G. Martinet. 2002. Joints, mortiers de pose et produits de ragréage: Les différentes pathologies: Réflexions et préconisations. *Pierre actual: Matériaux, ouvrages, techniques* (785): 66–79.

Bromblet, P., and V. Vergès-Belmin. 1996. L'élimination des sulfates sur la statuaire calcaire de plein air: Une habitude discutable. In *Le dessalement des matériaux poreux: 7es Journées d'études de la SFIIC, Poitiers, 9–10 mai 1996*, 55–64. Champs-sur-Marne: SFIIC.

Bromblet, P., M. Labouré, and G. Orial. 2003. Diversity of the cleaning procedures including laser for the restoration of carved portals in France over the last 10 years. *Journal of Cultural Heritage* 4 (Suppl. 1: Lasers in the Conservation of Artworks – LACONA IV): 17–26.

Bromblet, P., J.-D. Mertz, V. Vergès-Belmin, and L. Leroux. 2002. Consolidation et hydrofugation de la pierre. *Monumental: Revue scientifique et technique de la Sous-direction des monuments historiques* (2002): 200–243.

Brunet, J., E. Guillamet, and J. Plassard. 1997. The elimination of graffiti at Rouffignac. *INORA: International Newsletter on Rock Art* (17): 11–15.

Brunet, J., P. Malaurent, and R. Lastennet. 2006. Évolution de l'état hydrique d'une paroi de la Salle des Taureaux de la grotte de Lascaux: Conséquences pour la conservation. *Studies in Conservation* 51 (4): 241–51.

Brunet, J., J. Vouvé, and P. Malaurent. 2000. Re-establishing an underground climate appropriate for the conservation of the prehistoric paintings and engravings at Lascaux. *Conservation and Management of Archaeological Sites* 4 (1): 33–45.

Bugani, S., M. Camaiti, L. Morselli, E. Van de Casteele, P. Cloetens, and K. Janssens. 2008. X-ray computed tomography as a non-destructive tool for stone conservation. In *9th International Conference on Non-Destructive Investigations and Microanalysis for the Diagnostics and Conservation of Cultural and Environmental Heritage: Art 2008, Jerusalem, May 25–30, 2008*. Jerusalem: ISAS International Seminars. http://www.ndt.net/article/art2008/papers/161Bugani.pdf.

Building Effects Review Group (BERG), Great Britain. 1989. *The Effects of Acid Deposition on Buildings and Building Materials in the United Kingdom*. London: H.M.S.O.

Butlin, R. N. 1991. Effects of air pollution on buildings and materials. *Proceeedings of the Royal Society of Edinburgh* 976: 255–72.

Butlin, R. N., A. T. Coote, M. Devenish, I. F. C. Hughes, C. M. Hutchens, J. G. Irwin, G. O. Lloyd, S. W. Massey, A. H. Webb, and T. J. S. Yates. 1992. A four-year study of stone decay in different pollution climates in the United Kingdom. In *Proceedings of the 7th International Congress on Deterioration and Conservation of Stone, Lisbon, 15–18 June 1992*, ed. J. Delgado Rodrigues, F. Henriques, and F. Telmo Jeremias, 345–53. Lisbon: Laboratório Nacional de Engenharia Civil.

Cabreroravel, L. 1993. Book review: La conservation de la pierre monumentale en France [The Conservation of Stone Monuments in France]. *Bulletin Monumental* 151 (3): 249–50.

Calia, A., M. Lettieri, G. Quarta, M. Laurenzi Tabasso, and A. M. Mecchi. 2004. Documentation and assessment of the most important conservation treatments carried out on Lecce stone monuments in the last two decades. In *Proceedings of the 10th International Congress on Deterioration and Conservation of Stone, Stockholm, June 27–July 2, 2004*, ed. D. Kwiatkowski and R. Löfvendahl, vol. 1, 431–38. Stockholm: ICOMOS Sweden.

Camaiti, M., S. Bugani, E. Bernardi, L. Morselli, and M. Matteini. 2007. Effects of atmospheric NO_x on biocalcarenite coated with different conservation products. *Applied Geochemistry* 22 (6): 1248–54.

Camuffo, D., M. Del Monte, C. Sabbioni, and O. Vittori. 1982. Wetting, deterioration and visual features of stone surfaces in an urban area. *Atmospheric Environment* 16 (9): 2253–59.

Canals i Salomó, A., M. García Diez, I. Sauceda Pizarro, and E. Carbonell Roura. 2005. Actividad arqueológica y conservación del arte ruprestre en la Cueva de Maltravieso (Cáceres, España). Special issue, *PH: Boletín del Instituto Andaluz del Patrimonio Histórico* 13 (53): 44–57.

Cañaveras, J. C., S. Sanchez-Moral, V. Soler, and C. Saiz-Jimenez. 2001. Microorganisms and microbially induced fabrics in cave walls. *Geomicrobiology Journal* 18 (3): 223–40.

Candoré, J. C., J. L. Bodnar, V. Detalle, and P. Grossel. 2008. Non-destructive testing of mural paintings, in situ, by infra-red photothermal radiometry. In *In Situ Monitoring of Monumental Surfaces: Proceedings of the International Workshop SMW08, 27–29 October 2008, Florence, Italy*, ed. P. Tiano and C. Pardini, 21–30. Florence: Edifir.

Caner Saltik, E., ed. 2007. *QuarryScapes Report: Inventory of Ancient Quarry Landscapes in Turkey; Their Characteristics, Production and State of Conservation*. Trondheim, Norway: QuarryScapes Project. http://www.quarryscapes.no/text/publications/QS_del2_report.pdf.

Caneva, G. 1993. Ecological approach to the genesis of calcium oxalate patinas on stone monuments. *Aerobiologia* 9 (2–3): 149–56.

Caneva, G., R. Gasperini, and O. Salvadori. 2008. Endolithic colonization of stone in six Jewish cemeteries in urban environments. In *Proceedings of the 11th International Congress on Deterioration and Conservation of Stone, 15–20 September 2008, Toruń, Poland*, ed. J. W. Lukaszewicz and P. Niemcewicz, 49–56. Toruń, Poland: Nicolaus Copernicus University.

Caneva, G., M. P. Nugari, and O. Salvadori. 1991. *Biology in the Conservation of Works of Art*. Rome: ICCROM.

Caneva, G., M. P. Nugari, and O. Salvadori, eds. 2008. *Plant Biology for Cultural Heritage: Biodeterioration and Conservation*. Los Angeles: Getty Conservation Institute.

Caneva, G., O. Salvadori, S. Ricci, and S. Ceschin. 2005. Ecological analysis and biodeterioration processes over time at the Hieroglyphic Stairway in the Copán (Honduras) archaeological site. *Plant Biosystems* 139 (3): 295–310.

Cappitelli, F., and C. Sorlini. 2008. Microorganisms attack synthetic polymers in items representing our cultural heritage. *Applied and Environmental Microbiology* 74 (3): 564–69.

Cappitelli, F., J. D. Nosanchuk, A. Casadevall, L. Toniolo, L. Brusetti, S. Florio, P. Principi, S. Borin, and C. Sorlini. 2007a. Synthetic consolidants attacked by melanin-producing fungi: Case study of the biodeterioration of Milan (Italy) Cathedral marble treated with acrylics. *Applied and Environmental Microbiology* 73 (1): 271–77.

Cappitelli, F., L. Toniolo, A. Sansonetti, D. Gulotta, G. Ranalli, E. Zanardini, and C. Sorlini. 2007b. Advantages of using microbial technology over traditional chemical technology in removal of black crusts from stone surfaces of historical monuments. *Applied and Environmental Microbiology* 73 (17): 5671–75.

Cappitelli, F., E. Zanardini, P. Principi, M. Realini, and C. Sorlini. 2002. The biodegradability of synthetic resins used as binding media in housepaints. In *Plastics in Art: History, Technology, Preservation*, ed. T. van Oosten, Y. Shashoua, and F. Waentig, 124–33. Kölner Beiträge zur Restaurierung und Konservierung von Kunst- und Kulturgut 15. Munich: Siegl.

Cardiano, P., P. Mineo, S. Sergi, R. C. Ponterio, M. Triscari, and P. Piraino. 2003. Epoxy-silica polymers as restoration materials: Part II. *Polymer* 44 (16): 4435–41.

Cardiano, P., R. C. Ponterio, S. Sergi, S. Lo Schiavo, and P. Piraino. 2005. Epoxy-silica polymers as stone conservation materials. *Polymer* 46 (6): 1857–64.

Cariati, F., L. Rampazzi, L. Toniolo, and A. Pozzi. 2000. Calcium oxalate films on stone surfaces: Experimental assessment of the chemical formation. *Studies in Conservation* 45 (3): 180–88.

Casoli, A., and G. Palla. 2002. Presence and racemisation of amino acids in calcium oxalate patinas: A case study from the baptistery in Parma, Italy. *Annali di Chimica* 92 (11–12): 1057–65.

Cassar, J. 2004. Composition and property data of Malta's building stone for the construction of a database. In *Architectural and Sculptural Stone in the Cultural Landscape,* ed. R. Přikryl and P. Siegl, 11–28. Prague: Charles University, Karolinum Press.

Cassar, J., A. Marrocchi, M. L. Santarelli, and M. Muscat. 2008. Control de los daños por cristalización en la caliza de Malta mediante inhibidores de sales = Controlling crystallization damage by the use of salt inhibitors on Malta's limestone. *Materiales de construccion* 58 (289–90): 281–93.

Cassar, M. 2005. *Climate Change and the Historic Environment.* London: Centre for Sustainable Heritage, University College London. http://eprints.ucl.ac.uk/2082/.

Casta, Lucien. 1988. Nature pétrologique des statues-menhirs de corse, processus d'altération et de dégradation. In *Bulletin de la Société des sciences historiques et naturelles de la Corse: Actes du colloque, Les statues-menhirs de Corse, Intérêt archéologique, problème de mise en valeur, Sartène, 9–10–11 octobre 1985,* ed. Groupement régional des sociétés archéologiques de la Corse, 93–108. Ajaccio: Groupement régional des sociétés archéologiques de la Corse.

Castanier, S., G. Le Métayer-Levrel, G. Orial, J.-F. Loubière, and J.-P. Perthuisot. 2000. Bacterial carbonatogenesis and applications to preservation and restoration of historic property. In *Of Microbes and Art: The Role of Microbial Communities in the Degradation and Protection of Cultural Heritage,* ed. O. Ciferri, P. Tiano, and G. Mastromei, 203–18. New York: Kluwer Academic/Plenum Publishers.

Castellani, F. 2005. Historical monuments: The film crew. *Nature* 433 (7022): 100–101. DOI:10.1038/433100a.

Castellini, P., E. Esposito, B. Marchetti, N. Paone, and E. P. Tomasini. 2003. New applications of scanning laser doppler vibrometry (SLDV) to non-destructive diagnostics of artworks: Mosaics, ceramics, inlaid wood and easel painting. *Journal of Cultural Heritage* 4 (Suppl. 1: Lasers in the Conservation of Artworks—LACONA IV): 321–29.

Castelvetro, V., C. De Vita, G. Giannini, D. Scalarone, and O. Chiantore. 2004. Waterborne organic and hybrid nanostructured polymer particles as film forming materials for the consolidation and protection of porous substrates. *Science and Technology for Cultural Heritage* 13 (1–2): 37–46.

Cavalletti, R., L. Lazzarini, L. Marchesini, and G. Marinelli. 1985. A new type of epoxy resin for the structural consolidation of badly decayed stones. In *Ve Congrès international sur l'altération et la conservation de la Pierre, Lausanne, 25–27.9.1985: Actes = 5th International Congress on Deterioration and Conservation of Stone, Lausanne, 25–27.9.1985: Proceedings,* ed. G. Félix, 769–78. Lausanne: Presses polytechniques romandes.

Cerveny, N. V. 2005. A weathering-based perspective on rock art conservation. PhD diss., Arizona State University.

Chamay, J. 1992. Discours de clôture: Quelques mots en guise de conclusion. In *La conservation des monuments dans le bassin méditerranéen: Actes du 2ème Symposium international = The Conservation of Monuments in the Mediterranean Basin: Proceedings of the 2nd International Symposium,* ed. D. Decrouez, J. Chamay, and F. Zezza, 518–19. Geneva: Ville de Genève, Muséum d'histoire naturelle, and Musée d'art et d'histoire.

Chandler, J. H., P. Bryan, and J. G. Fryer. 2007. The development and application of a simple methodology for recording rock art using consumer-grade digital cameras. *Photogrammetric Record* 22 (117): 10–21.

Chandler, J. H., J. G. Fryer, and H. T. Kniest. 2005. Non-invasive three-dimensional recording of Aboriginal rock art using cost-effective digital photogrammetry. *Rock Art Research: The Journal of the Australian Rock Art Research Association (AURA)* 22 (2): 119–30.

Charola, A. E. 1995. Water-repellent treatments for building stones: A practical overview. *APT Bulletin* 26 (2–3): 10–17.

Charola, A. E. 2000. Salts in the deterioration of porous materials: An overview. *Journal of the American Institute for Conservation* 39 (3): 327–43.

Charola, A. E., and R. Ware. 2002. Acid deposition and the deterioration of stone: A brief review of a broad topic. In *Natural Stone, Weathering Phenomena, Conservation Strategies and Case Studies,* ed. S. Siegesmund, T. Weiss, and A. Vollbrecht, 393–406. Geological Society Special Publications 205. London: Geological Society of London.

Charola, A. E., J. Weber, and E. Bolle. 1990. Preliminary studies of deterioration patterns at Cerro Colorado, Argentina: Open cave formation and honeycomb weathering. In *9th Triennial Meeting, ICOM Committee for Conservation, Dresden, German Democratic Republic, 26–31 August 1990: Preprints,* ed. K. Grimstad, 321–25. Los Angeles: ICOM Committee for Conservation; Marina del Rey, CA: Getty Conservation Institute.

Chiantore, O., and M. Lazzari. 2001. Photo-oxidative stability of paraloid acrylic protective polymers. *Polymer* 42 (1): 17–27.

Chiari, G., and R. Cossio. 2004. Lichens on Wyoming sandstone: Do they cause damage? In *Biodeterioration of Stone Surfaces: Lichens and Biofilms as Weathering Agents of Rocks and Cultural Heritage,* ed. L. L. St. Clair and M. R. D. Seaward, 99–114. Dordrecht, Netherlands, and London: Kluwer Academic Publishers.

Chiari, G., and M. Leona. 2005. The state of conservation science. *Conservation: The GCI Newsletter* 20 (2), 4–9. http://www.getty.edu/conservation/publications/newsletters/pdf/v20n2.pdf.

Ciabach, J. 1996. The effect of water soluble salts on the impregnation of sandstone with silicone microemulsions. In *Proceedings of the 8th International Congress on Deterioration and Conservation of Stone, Berlin, 30 Sept.–4 Oct. 1996,* ed. J. Riederer, 1215–21. Berlin: Möller Druck und Verlag.

Ciferri, O., P. Tiano, and G. Mastromei, eds. 2000. *Of Microbes and Art: The Role of Microbial Communities in the Degradation and Protection of Cultural Heritage.* New York: Kluwer Academic/Plenum Publishers.

Clarke, B. L., and J. Ashurst. 1972. *Stone Preservation Experiments.* Watford: Building Research Establishment; London: Directorate of Ancient Monuments and Special Services.

Clottes, J. 2006. Rock art today. *Conservation: The GCI Newsletter* 21 (3). http://www.getty.edu/conservation/publications/newsletters/21_3/feature.html.

Cnudde, V., M. Dierick, J. Vlassenbroeck, B. Masschaele, E. Lehmann, P. Jacobs, and L. Van Hoorebeke. 2008. High-Speed neutron radiography for monitoring the water absorption by capillarity in porous materials. *Nuclear Inst. and Methods in Physics Research, B* 266 (1): 155–163.

Cnudde, V., P. Dubruel, K. De Winne, I. De Witte, B. Masschaele, P. Jacobs, and E. Schacht. 2009. The use of X-ray tomography in the study of water repellents and consolidants. *Engineering Geology* 103 (3–4): 84–92.

Coe, J. A., S. I. Sherwood, J. A. Messerich, C. L. Pillmore, A. Andersen, and V. G. Mossotti. 1992. Measuring stone decay with close range photogrammetry. In *Proceedings of the 7th International Congress on Deterioration and Conservation of Stone, Lisbon, 15–18 June 1992,* ed. J. Delgado Rodrigues, F. Henriques, and F. Telmo Jeremias, 917–26. Lisbon: Laboratório Nacional de Engenharia Civil.

Coffman, R. L., N. Agnew, and C. Selwitz. 1991. Modification of the physical properties of natural and artificial adobe by chemical consolidation. In *Materials Issues in Art and Archaeology II Symposium, Held April 17–21, 1990, San Francisco,* ed. P. B. Vandiver, J. R. Druzik, and G. Segan Wheeler, 201–7. Materials Research Society Symposium Proceedings 185. Pittsburgh: Materials Research Society.

Colombo, C., C. Daffara, R. Fontana, M. Ch. Gambino, M. Mastroianni, E. Pampaloni, M. Realini, and A. Sansonetti. 2007. Evaluation by lasermicroprofilometry of morphological changes induced on stone materials by laser cleaning. In *Lasers in the Conservation of Artworks: LACONA VI, Vienna, Austria, Sept. 21–25, 2005; Proceedings,* ed. J. Nimmrichter, W. Kautek, and M. Schreiner, 523–26. Springer Proceedings in Physics 116. Berlin and New York: Springer.

Cooke, R. U., and G. B. Gibbs. 1993. *Crumbling Heritage? Studies of Stone Weathering in Polluted Atmospheres: A Report of Research on Atmospheric Pollution and Stone Decay for the Joint Working Party Between the Cathedrals Fabric Commission for England and the Joint Environmental Programme of National Power Plc and Powergen Plc.* London: University College, Department of Geography.

Cooper, M., ed. 1998. *Laser Cleaning in Conservation: An Introduction.* Woburn, MA: Butterworth-Heinemann.

Cooper, M. I., D. C. Emmony, and J. H. Larson. 1993. The evaluation of laser cleaning of stone sculpture. In *Structural Repair and Maintenance of Historical Buildings III,* ed. C. A. Brebbia and R. J. B. Frewer, 259–66. Southampton and Boston: Computational Mechanics Publications.

Cooper, M. I., D. C. Emmony, and J. Larson. 1995. Characterization of laser cleaning of limestone. *Optics and Laser Technology* 27 (1): 69–73.

Correia, J. M. 2005. Treatment evaluation for the micritic limestone at Mission San José y San Miguel de Aguayo, San Antonio, Texas. Master's thesis, University of Pennsylvania.

Correia, J., and F. Matero. 2008. Calcium tartrate tetrahydrate preconsolidation of salt-contaminated limestone at Mission San José y San Miguel de Aguayo. *Journal of the American Institute for Conservation* 47 (2): 81–95.

Costa, D., and J. Delgado Rodrigues. 2008a. Consolidation treatment of salt laden materials: Methodology for their laboratory study. In *Proceedings of the 11th International Congress on Deterioration and Conservation of Stone, 15–20 September 2008, Toruń, Poland,* ed. J. W. Lukaszewicz and P. Niemcewicz, 827–35. Toruń, Poland: Nicolaus Copernicus University.

Costa, D., and J. Delgado Rodrigues. 2008b. Evaluation of consolidation treatments applied to granitic materials: Experience and critical overview of laboratory testing. In *Stone Consolidation in Cultural Heritage: Research and Practice; Proceedings of the International Symposium, Lisbon, 6–7 May 2008,* ed. J. Delgado Rodrigues and J. M. Mimoso, 389–98. Lisbon: LNEC (Laboratório Nacional de Engenharia Civil).

Crispim, C. A., and C. C. Gaylarde. 2005. Cyanobacteria and biodeterioration of cultural heritage: A review. *Microbial Ecology* 49 (1): 1–9.

Croci, G., and J. Delgado Rodrigues. 2002. Surface and structural stability for the conservation of historic buildings. In *Science and Technology of the Environment for Sustainable Protection of Cultural Heritage: EC Advanced Study Course.* London: University College London. http://www.ucl.ac.uk/sustainableheritage/Archive_0906/sustainableheritage/sustainableheritage/learning/asc/delegates/coursematerials.html.

Croveri, P., and M. Chiavarini. 2000. Evaluations of the effectiveness of innovative perfluoropolyurethanes polymers as consolidants for porous materials. In *Proceedings of the 9th International Congress on Deterioration and Conservation of Stone, Venice, June 19–24, 2000*, ed. V. Fassina, vol. 2, 263–71. Amsterdam and New York: Elsevier.

Cuzman, O. A., P. Tiano, and S. Ventura. 2008. New control methods against biofilms' formation on the monumental stones. In *Proceedings of the 11th International Congress on Deterioration and Conservation of Stone, 15–20 September 2008, Toruń, Poland,* ed. J. W. Lukaszewicz and P. Niemcewicz, 837–46. Toruń, Poland: Nicolaus Copernicus University.

Dajnowski, A., A. Jenkins, and A. Lins. 2009. The use of lasers for cleaning large architectural structures. *APT Bulletin* 40 (1): 13–23.

Dal Bianco, B., and R. Bertoncello. 2008. Sol-gel silica coatings for the protection of cultural heritage glass. *Nuclear Instruments and Methods in Physics Research, Section B: Beam Interactions with Materials and Atoms* 266 (10): 2358–62.

Dandridge, D. E. 2000. Overview of conservation treatments applied to rock glyph archaeological sites. Master's thesis, Texas A&M University.

Dandridge, D. E. 2006. Lichens: The challenge for rock art conservation. PhD diss., Texas A&M University.

Daniele, V., and G. Taglieri. 2010. Nanolime suspensions applied on natural lithotypes: The influence of concentration and residual water content on carbonatation process and on treatment effectiveness. *Journal of Cultural Heritage* 11 (1): 102–6.

Davis, K. J., and A. Lüttge. 2005. Quantifying the relationship between microbial attachment and mineral surface dynamics using vertical scanning interferometry (VSI). Special issue, *American Journal of Science* 305 (6–8): 727–51.

Deacon, J. 2006. Rock art conservation and tourism. *Journal of Archaeological Method and Theory* 13 (4): 379–99.

Dean, J. C. 1997. The Kokopelli dilemma: The use, abuse, and care of rock art. *Newsletter: Western Association for Art Conservation* 19 (3): 18–25. http://cool.conservation-us.org/waac/wn/wn19/wn19-3/wn19-309.html.

Dean, J. C. 1998. Bring conservation to the mountain, not the mountain to the conservator: The care of rock art sites in the USA. In *Site Effects: The Impact of Location on Conservation Treatments,* ed. M. M. Wright and Y. M. T. Player-Dahnsjö, 98–102. Edinburgh: Scottish Society for Conservation & Restoration.

Dean, J. C. 2001. Rock image conservation at Petroglyph National Monuments. *CRM* 24 (6): 17. http://crm.cr.nps.gov/issue.cfm?volume=24&number=06.

Dean, J. C., J. Flood, J. Van Tilburg, N. Agnew, and J. Levin. 2006. Preserving a worldwide heritage: A discussion about rock art conservation. *Conservation: The GCI Newsletter* 21 (3). http://www.getty.edu/conservation/publications/newsletters/21_3/dialogue.html.

De Clercq, H. 2008. Performance of limestone contaminated with binary mixtures of sodium sulphate and treated with a water repellent. In *Hydrophobe V: Water Repellent Treatment of Building Materials; Proceedings of Hydrophobe V, Fifth International Conference on Water Repellent Treatment of Building Materials, Royal Institute for Cultural Heritage (KIK-IRPA), Brussels, Belgium, April 15 and 16, 2008,* ed. H. De Clercq and A. E. Charola, 107–15. Freiburg: Aedificatio Verlag.

De Clercq, H. and A. E. Charola, ed. 2008. *Hydrophobe V: Water Repellent Treatment of Building Materials: Proceedings of Hydrophobe V, Fifth International Conference on Water Repellent Treatment of Building Materials, Royal Institute for Cultural Heritage (KIK-IRPA), Brussels, Belgium, April 15th and 16th, 2008.* Freiburg: Aedificatio Publishers.

De Clercq, H., and E. De Witte. 2001. Effectiveness of silicon based water repellent agents at different application conditions, Part II: Commercial water repellents. *Restoration of Buildings and Monuments: An International Journal – Bauinstandsetzen und Baudenkmalpflege: Eine internationale Zeitschrift* 7 (6): 641–54.

De Clercq, H., S. De Zanche, and G. Biscontin. 2007. TEOS and time: The influence of application schedules on the effectiveness of ethyl silicate based consolidants. *Restoration of Buildings and Monuments: An International Journal = Bauinstandsetzen und Baudenkmalpflege: Eine internationale Zeitschrift* 13 (5): 305–18. http://www.icvbc.cnr.it/drilling/publications/032_DE%20CLERCQ.pdf.

DeCruz, A., M. L. Wolbarsht, A. Andreotti, M. P. Colombini, D. Pinna, and C. F. Culberson. 2009. Investigation of the Er: YAG laser at 2.94 µm to remove lichens growing on stone. *Studies in Conservation* 54 (4): 268–77.

Degryse, P., E. Bloxam, T. Heldal, P. Storemyr, and M. Waelkens. 2006. Quarries in the landscape: A survey of the area of Sagalassos (SW Turkey). In *Conservation of Ancient Stone Quarry Landscapes in the Eastern Mediterranean: First QuarryScapes Symposium, October 15–17, 2006, Antalya, Turkey: Extended Abstract Collection,* ed. P. Degryse, 7–10. Trondheim, Norway: QuarryScapes Project. http://www.quarryscapes.no/text/Workshops/Proceedings_1stQS_symposium.pdf#page=7.

Dei, L., and B. Salvadori. 2006. Nanotechnology in cultural heritage conservation: Nanometric slaked lime saves architectonic and artistic surfaces from decay. *Journal of Cultural Heritage* 7 (2): 110–15.

Delegou, E. T., N. P. Avdelidis, E. Karaviti, and A. Moropoulou. 2008. NDT&E techniques and SEM-EDS for the assessment of cleaning interventions on Pentelic marble surfaces. *X-Ray Spectrometry* 37 (4): 435–43.

Delgado Rodrigues, J. 2001. Evaluaciön del comportamiento expansivo de las rocas y su interés en conservación = Swelling behaviour of stones and its interest in conservation. An appraisal. *Materiales de construccion* 2001 (263–64): 183–95.

Delgado Rodrigues, J., and J. Valero. 2003. A brief note on the elimination of dark stains of biological origin. *Studies in Conservation* 48 (1): 17–22.

Delgado Rodrigues, J., A. Ferreira Pinto, and D. Rodrigues da Costa. 2002. Tracing of decay profiles and evaluation of stone treatments by means of microdrilling techniques. *Journal of Cultural Heritage* 3 (2): 117–25.

Del Monte, M. 1992. Stone monument decay and air pollution. In *Weathering and Air Pollution: Primo corso della Scuola universitaria C.U.M. conservazione dei monumenti, Lago di Garda (Portese), Venezia, Milano, 2–9 settembre 1991*, ed. F. Zezza, 101–10. Bari: Mario Adda Editore.

Del Monte, M., and C. Sabbioni. 1987. A study of the patina called *scialbatura* on imperial Roman marbles. *Studies in Conservation* 32 (3): 114–21.

Demas, M. 2004. "Site unseen": The case for reburial of archaeological sites. *Conservation and Management of Archaeological Sites* 6 (3–4): 137–54.

De Muynck, W., N. De Belie, and W. Verstraete. 2010. Microbial carbonate precipitation in construction materials: A review. *Ecological Engineering* 36 (2): 118–36.

De Oliveira Castello Branco, H. D., and L. A. Cruz Souza. 2002. Rock art conservation in the Peruaçu valley, Minas Gerais, Brazil. In *13th Triennial Meeting, ICOM Committee for Conservation, Rio de Janiero, 22–27 September 2002: Preprints*, ed. R. Vontobel, 556–59. London: James & James (Science Publishers) Ltd.

De Witte, E., ed. 2001. *Salt Compatibility of Surface Treatments (SCOST): Final Report*. European Contract ENV4-CT98-0710. Brussels: KIK-IRPA.

De Witte, E., and M. Dupas. 1992. Cleaning poultices based on EDTA. In *Proceedings of the 7th International Congress on Deterioration and Conservation of Stone, Lisbon, 15–18 June 1992*, ed. J. Delgado Rodrigues, F. Henriques, and F. Telmo Jeremias, 1023–31. Lisbon: Laboratório Nacional de Engenharia Civil.

Diakumaku, E., A. A. Gorbushina, W. E. Krumbein, L. Panina, and S. Soukharjevski. 1995. Black fungi in marble and limestones: An aesthetical, chemical and physical problem for the conservation of monuments. *Science of the Total Environment* 167: 295–304.

Díaz-Andreu, M., C. Brooke, M. Rainsbury, and Nick Rosser. 2006. The spiral that vanished: The application of non-contact recording techniques to an elusive rock art motif at Castlerigg Stone Circle in Cumbria. *Journal of Archaeological Science* 33 (11): 1580–87.

Dingelstadt, C., R. Dreesen, J. Thorez, G. Lorenzi, D. Bossiroy, D. Antenucci, and J. Banier. 2000. A petrographical atlas as a decision-tool in replacement and substitution of ornamental stones in historical buildings and monuments. In *Proceedings of the 9th International Congress on Deterioration and Conservation of Stone, Venice, June 19–24, 2000*, ed. V. Fassina, vol. 2, 33–41. Amsterdam and New York: Elsevier.

Doehne, E. 1994. In situ dynamics of sodium sulfate hydration and dehydration in stone pores: Observations at high magnification using the environmental scanning electron microscope. In *Conservation of Monuments in the Mediterranean Basin: Stone Monuments, Methodologies for the Analysis of Weathering and Conservation; Proceedings of the 3rd International Symposium, Venice, 22–25 June 1994 = La conservazione dei monumenti nel bacino del Mediterraneo: Materiali lapidei e monumenti, metodologie per l'analisi del degrado e la conservazione; Atti del 3° simposio internazionale, Venezia, 22–25 giugno 1994*, ed. V. Fassina, H. Ott, and F. Zezza, 143–50. Venice: Soprintendenza ai beni artistici e storici di Venezia.

Doehne, E. 2002. Salt weathering: A selective review. In *Natural Stone, Weathering Phenomena, Conservation Strategies and Case Studies*, ed. S. Siegesmund, T. Weiss, and A. Vollbrecht, 51–64. Geological Society Special Publications 205. London: Geological Society of London.

Doehne, E. 2003. Building material decay and salt weathering: A selected bibliography. Supplement to *Natural Stone, Weathering Phenomena, Conservation Strategies and Case Studies*, ed. S. Siegesmund, T. Weiss, and A. Vollbrecht. Geological Society Special Publication 205. London: Geological Society of London. www.geolsoc.org.uk/webdav/site/GSL/shared/Sup_pubs/2003/SUP18182.rtf.

Doehne, E., and J. L. Drever. 1994. Appendix I: Report on the impromptu discussion session on stone conservation. In *Durability and Change: The Science, Responsibility, and Cost of Sustaining Cultural Heritage; December 6–11, 1992*, ed. W. E. Krumbein, P. Brimblecombe, D. E. Cosgrove, and S. Staniforth, 287–89. Dahlem Workshop Reports: Environmental Sciences Research Report ES 15. Chichester and New York: John Wiley.

Doehne, E., and S. Pinchin. 2008. Time-lapse macro-imaging in the field: Monitoring rapid flaking of magnesian limestone. In *Proceedings of the 11th International Congress on Deterioration and Conservation of Stone, 15–20 September 2008, Toruń, Poland,* ed. J. W. Lukaszewicz and P. Niemcewicz, 365–72. Toruń, Poland: Nicolaus Copernicus University.

Doehne, E., M. Schiro, T. Roby, G. Chiari, G. Lambousy, and H. Knight. 2008. Evaluation of poultice desalination process at Madame Johns' Legacy, New Orleans. In *Proceedings of the 11th International Congress on Deterioration and Conservation of Stone, 15–20 September 2008, Toruń, Poland,* ed. J. W. Lukaszewicz and P. Niemcewicz, 857–64. Toruń, Poland: Nicolaus Copernicus University.

Doehne, E., S. Simon, U. Mueller, D. Carson, and A. Ormsbee. 2005. Characterization of carved rhyolite tuff: The Hieroglyphic Stairway of Copán, Honduras. *Restoration of Buildings and Monuments: An International Journal = Bauinstandsetzen und Baudenkmalpflege: Eine internationale Zeitschrift* 11 (4): 247–54.

Dogariu, A., V. Zafiropulos, P. Pouli, V. Kylikoglou, P. Maravelaki-Kalaitzaki, and B. S. Luk'yanchuk. 2005. Synchronous use of IR and UV laser pulses in the removal of encrustation: Mechanistic aspects, discoloration phenomena and benefits. In *Lasers in the Conservation of Artworks: LACONA V, Osnabrück, Germany, Sept. 15–18, 2003; Proceedings,* ed. K. Dickmann, C. Fotakis, and J. F. Asmus, 311–18. Springer Proceedings in Physics 100. Berlin: Springer.

Doherty, B., M. Pamplona, R. Selvaggi, C. Miliani, M. Matteini, A. Sgamellotti, and B. Brunetti. 2007. Efficiency and resistance of the artificial oxalate protection treatment on marble against chemical weathering. *Applied Surface Science* 253 (10): 4477–84.

Domaslowski, W. 1969. Consolidation of stone objects with epoxy resins. *Monumentum* 4: 51–64. http://www.international.icomos.org/monumentum/vol4/vol4_3.pdf.

Domaslowski, W., ed. 2003. *Preventive Conservation of Stone Historical Objects.* Torún, Poland: Nicolaus Copernicus University.

Domaslowski, W., and D. Sobkowiak. 1991. Ocena wlasciwosci kamieni wzmocnionych swiatlotrwala, cykloalifatyczna zywica epoksydowa Eurostac EP 2101 = Evaluation of properties of stones consolidated with light-resistant cycloaliphatic epoxy resin (Eurostatic EP 2101). *Acta universitatis nicolai copernici* 18 (227): 67–81.

Domaslowski, W., and A. Strzelczyk. 1986. Evaluation of applicability of epoxy resins to conservation of stone historic monuments. In *Case Studies in the Conservation of Stone and Wall Paintings: Preprints of the Contributions to the Bologna Congress, 21–26 September 1986,* ed. N. S Brommelle and P. Smith, 126–32. London: International Institute for Conservation of Historic and Artistic Works.

Dorn, R. I., D. S. Whitley, N. V. Cerveny, S. J. Gordon, C. D. Allen, and E. Gutbrod. 2008. The Rock Art Stability Index: A new strategy for maximizing the sustainability of rock art. *Heritage Management* 1 (1): 37–70.

Dornieden, Th., A. A. Gorbushina, and W. E. Krumbein. 2000. Biodecay of cultural heritage as a space/time-related ecological situation: An evaluation of a series of studies. *International Biodeterioration and Biodegradation* 46 (4): 261–70.

Dragovich, D. 1981. Cavern microclimates in relation to preservation of rock art. *Studies in Conservation* 26 (4): 143–49.

Drdácký, M., Z. Slížková, and G. Ziegenbalg. 2009. A nano approach to consolidation of degraded historic lime mortars. *Journal of Nano Research* 8: 13–22.

Duffus, P., T. Wangler, and G. W. Scherer. 2008. Swelling damage mechanism for clay-bearing sandstones. In *Proceedings of the 11th International Congress on Deterioration and Conservation of Stone, 15–20 September 2008, Toruń, Poland,* ed. J. W. Lukaszewicz and P. Niemcewicz, 65–72. Toruń, Poland: Nicolaus Copernicus University.

Dupont, J., C. Jacquet, B. Dennetière, S. Lacoste, F. Bousta, G. Orial, C. Cruaud, A. Couloux, and M.-F. Roquebert. 2007. Invasion of the French Paleolithic painted cave of Lascaux by members of the *Fusarium solani* species complex. *Mycologia* 99 (4): 526–33.

D'Urbano, M. S., C. Giovannone, P. Governale, A. Pandolfi, and U. Santamaria. 1994. La pulitura laser di superfici lapidee: Messa a punto di una metodologia standardizzata per il controllo degli effetti = A standardized methodology to check the effects of laser cleaning of stone surfaces. In *Conservation of Monuments in the Mediterranean Basin: Stone Monuments, Methodologies for the Analysis of Weathering and Conservation; Proceedings of the 3rd International Symposium, Venice, 22–25 June 1994 = La conservazione dei monumenti nel bacino del Mediterraneo: Materiali lapidei e monumenti, metodologie per l'analisi del degrado e la conservazione; Atti del 3° simposio internazionale, Venezia, 22–25 giugno 1994*, ed. V. Fassina, H. Ott, and F. Zezza, 955–62. Venice: Soprintendenza ai beni artistici e storici di Venezia.

Duthie, L., E. Hyslop, C. Kennedy, V. Phoenix, and M. R. Lee. 2008. Quantitative assessment of decay mechanisms in Scottish building sandstones. In *Proceedings of the 11th International Congress on Deterioration and Conservation of Stone, 15–20 September 2008, Toruń, Poland*, ed. J. W. Lukaszewicz and P. Niemcewicz, 73–80. Toruń, Poland: Nicolaus Copernicus University.

Egleston, T. 1886. The cause and prevention of the decay of building stone. *Transactions of the American Society of Civil Engineers* 15: 341.

El-Hakim, S., J. Fryer, M. Picard, and E. Whiting. 2004. Digital recording of Aboriginal rock art. In *Proceedings of the Tenth International Conference on Virtual Systems and Multimedia (VSMM 2004): Hybrid Realities & Digital Partners; Explorations in Art, Heritage, Science & the Human Factor, 17–19 November 2004, Ogaki City, Japan*, ed. H. Thwaites, 344–53. Amsterdam and Fairfax, VA: IOS Press.

Elmer, K., E. Rose, B. Fitzner, Wolfgang E. Krumbein, and T. Warscheid. 1993. Sterilisation of cultural objects by ethylene oxide: Application for conservation practice by laboratory-based or mobile treatments. In *Conservation of Stone and Other Materials: Proceedings of the International RILEM/UNESCO Congress "Conservation of Stone and Other Materials: Research—Industry—Media," UNESCO Headquarters, Paris, June 29–July 1, 1993*, ed. M.-J. Thiel, 581–88. RILEM Proceedings 21. London and New York: E & FN Spon.

English Heritage. 2009. Strategic Stone Survey. Available at http://www.english-heritage.org.uk/server/show/nav.21370.

English Stone Forum. 2009. English Heritage Strategic Stone Study, Update March 2009. Available at http://www.englishstone.org.uk/sss%203-2009.html.

Escalante, M. R., J. Valenza, and G. W. Scherer. 2000. Compatible consolidants from particle-modified gels. In *Proceedings of the 9th International Congress on Deterioration and Conservation of Stone, Venice, June 19–24, 2000*, ed. V. Fassina, vol. 2, 459–65. Amsterdam and New York: Elsevier.

Espinosa, R. M., L. Franke, and G. Deckelmann. 2008. Model for the mechanical stress due to the salt crystallization in porous materials. *Construction and Building Materials* 22 (7): 1350–67.

Espinosa Marzal, R. M., and G. W. Scherer. 2008a. Crystallization of sodium sulfate salts in limestone. *Environmental Geology* 56 (3–4): 605–21.

Espinosa Marzal, R. M., and G. W. Scherer. 2008b. Study of the pore clogging induced by salt crystallization in Indiana limestone. In *Proceedings of the 11th International Congress on Deterioration and Conservation of Stone, 15–20 September 2008, Toruń, Poland*, ed. J. W. Lukaszewicz and P. Niemcewicz, 81–88. Toruń, Poland: Nicolaus Copernicus University.

Espinosa Marzal, R. M., and G. W. Scherer. 2009. Crystallization pressure exerted by in-pore confined crystals. In *Poromechanics IV: Proceedings of the Fourth Biot Conference on Poromechanics*, ed. Hoe I. Ling, A. Smyth, and R. Betti, 1013–18. Lancaster, PA: Destech Publications Inc.

Espinosa Marzal, R. M., and G. W. Scherer. Forthcoming. Damaging mechanisms of salt crystallization. In *Limestone in the Built Environment: Present Day Challenges for Preservation of the Past*, ed. B. J. Smith, M. Gomez Heras, H. A. Viles, and J. Cassar. Geological Society of London Special Publication. London: Geological Society of London.

European Union (EU). 2008. Directive 2008/50/EC of the European Parliament and of the Council of 21 May 2008 on ambient air quality and cleaner air for Europe. *Official Journal of the European Union* L152, 11.6.2008: 33. http://eur-lex.europa.eu/LexUriServ/LexUriServ.do?uri=OJ:L:2008:152:0001:0044:EN:PDF.

Facaoaru, I., and C. Lugnani. 1993. Contributions to the diagnosis of stone and concrete historical structures using nondestructive techniques. In *Conservation of Stone and Other Materials: Proceedings of the International RILEM/UNESCO Congress "Conservation of Stone and Other Materials: Research—Industry—Media," UNESCO Headquarters, Paris, June 29–July 1, 1993*, ed. M.-J. Thiel, 238–51. RILEM Proceedings 21. London and New York: E & FN Spon.

Fassina, V. 1992. Atmospheric pollutants responsible for stone decay: Wet and dry surface deposition of air pollutants on stone and the formation of black scabs. In *Weathering and Air Pollution: Primo corso della Scuola universitaria C.U.M. conservazione dei monumenti, Lago di Garda (Portese), Venezia, Milano, 2–9 settembre 1991*, ed. F. Zezza, 67–86. Bari: Mario Adda Editore.

Fassina, V. 1994. General criteria for the cleaning of stone: Theoretical aspects and methodology of application. In *Stone Material in Monuments: Diagnosis and Conservation; Scuola universitaria C.U.M. conservazione dei monumenti, Heraklion, Crete, 24–30 May 1993*, ed. F. Zezza, 131–38. Bari: Mario Adda Editore.

Fassina, V. 2008. European Technical Committee 346: Conservation of cultural property: Updating of the activity after a three year period. In *9th International Conference on Non-Destructive Investigations and Microanalysis for the Diagnostics and Conservation of Cultural and Environmental Heritage: Art 2008, Jerusalem, May 25–30, 2008*. Jerusalem: ISAS International Seminars. http://www.ndt.net/article/art2008/papers/174Fassina.pdf.

Fassina, V., G. Gaudini, and R. Cavaletti. 2008. Condition assessment of marble sculptures on polychromed and gilded decorations of the Arca Scaligera of Cansignorio della Scala in Verona. In *Proceedings of the 11th International Congress on Deterioration and Conservation of Stone, 15–20 September 2008, Toruń, Poland*, ed. J. W. Lukaszewicz and P. Niemcewicz, 89–96. Toruń, Poland: Nicolaus Copernicus University.

Fassina, V., R. Arbizzani, C. Botteghi, U. Matteoli, E. Passaglia, and M. Aglietto. 1994. Behaviour of 2,2,2-trifluoroethylmethacrylate polymers as stone protective materials. In *Conservation of Monuments in the Mediterranean Basin: Stone Monuments, Methodologies for the Analysis of Weathering and Conservation; Proceedings of the 3rd International Symposium, Venice, 22–25 June 1994 = La conservazione dei monumenti nel bacino del Mediterraneo: Materiali lapidei e monumenti, metodologie per l'analisi del degrado e la conservazione; Atti del 3° simposio internazionale, Venezia, 22–25 giugno 1994*, ed. V. Fassina, H. Ott, and F. Zezza, 911–23. Venice: Soprintendenza ai beni artistici e storici di Venezia.

Favaro, M., R. Mendichi, F. Ossola, U. Russo, S. Simon, P. Tomasin, and P. A. Vigato. 2006. Evaluation of polymers for conservation treatments of outdoor exposed stone monuments, Part I: Photo-oxidative weathering. *Polymer Degradation and Stability* 91 (12): 3083–96.

Favaro, M., R. Mendichi, F. Ossola, S. Simon, P. Tomasin, and P. A. Vigato. 2007. Evaluation of polymers for conservation treatments of outdoor exposed stone monuments, Part II: Photo-oxidative and salt-induced weathering of acrylic-silicone mixtures. *Polymer Degradation and Stability* 92 (3): 335–51.

Favaro, M., F. Ossola, P. Tomasin, P. A. Vigato, G. Rossetto, N. El Habra, and M. Casarin. 2008. A novel approach to compatible and durable consolidation of limestone. In *Proceedings of the 11th International Congress on Deterioration and Conservation of Stone, 15–20 September 2008, Toruń, Poland*, ed. J. W. Lukaszewicz and P. Niemcewicz, 865–72. Toruń, Poland: Nicolaus Copernicus University.

Favaro, M., S. Simon, C. Menichelli, V. Fassina, and P. A. Vigato. 2005. The four virtues of the Porta della Carta, Ducal Palace, Venice: Assessment of the state of preservation and re-evaluation of the 1979 restoration. *Studies in Conservation* 50 (2): 109–27.

Feilden, B. M. 2003. *Conservation of Historic Buildings*. 3rd ed. Oxford: Architectural Press.

Félix, C. 1996. Peut-on consolider les grès tendres du Plateau suisse avec le silicate d'éthyle? In *Conservation et restauration des biens culturels: Pierre, pollution atmosphérique, peinture murale, études scientifiques et cas pratiques; Actes du Congrès LCP 1995, Montreux 24–29 septembre 1995 = Preservation and Restoration of Cultural Heritage: Stone Materials, Air Pollution, Murals,*

Scientific Research Work and Case Studies: Proceedings of the 1995 LPC Congress, Montreux 24–29 September 1995, ed. R. Pancella, 267–74. Lausanne: Laboratoire de conservation de la pierre, Département des matériaux, EPFL (Ecole polytechnique fédérale de Lausanne).

Félix, C., and V. Furlan. 1994. Variations dimensionnelles de grès et calcaires liées à leur consolidation avec un silicate d'éthyle = Dimensional variations of sandstone and limestone in connection with their consolidation with ethyl silicate. In *Conservation of Monuments in the Mediterranean Basin: Stone Monuments: Methodologies for the Analysis of Weathering and Conservation: Proceedings of the 3rd International Symposium, Venice, 22–25 June 1994 = La conservazione dei monumenti nel bacino del Mediterraneo: Materiali lapidei e monumenti, metodologie per l'analisi del degrado e la conservazione: Atti del 3° simposio internazionale, Venezia, 22–25 giugno 1994,* ed. V. Fassina, H. Ott, and F. Zezza, 855–59. Venice: Soprintendenza ai beni artistici e storici di Venezia.

Ferreira Pinto, A., J. Delgado Rodrigues, S. Bracci, and B. Sacchi. 2008. The action of APTES as coupling agent of ethylsilicate for limestone and marble consolidation. In *Stone Consolidation in Cultural Heritage: Research and Practice; Proceedings of the International Symposium, Lisbon, 6–7 May 2008,* ed. J. Delgado Rodrigues and J. M. Mimoso, 71–80. Lisbon: LNEC (Laboratório Nacional de Engenharia Civil).

Ferreira Pinto, A. P., and J. Delgado Rodrigues. 2008. The action of inorganic consolidants in limestones. In *Proceedings of the 11th International Congress on Deterioration and Conservation of Stone, 15–20 September 2008, Toruń, Poland,* ed. J. W. Lukaszewicz and P. Niemcewicz, 873–80. Toruń, Poland: Nicolaus Copernicus University.

Ferreira Pinto, A. P., and J. Delgado Rodrigues. 2008. Hydroxylating conversion treatment and alkoxysilane coupling agent as pre-treatment for the consolidation of limestones with ethyl silicate. In *Stone Consolidation in Cultural Heritage: Research and Practice; Proceedings of the International Symposium, Lisbon, 6–7 May 2008,* ed. J. Delgado Rodrigues and J. M. Mimoso, 131–40. Lisbon: LNEC (Laboratório Nacional de Engenharia Civil).

Fidler, J. 1995. Lime treatments: Lime watering and shelter coating of friable historic masonry. *APT Bulletin* 26 (4): 50–56.

Fidler, J. 2002. Lime treatments: An overview of lime watering and shelter coating of friable historic limestone masonry. In *Stone: Stone Building Materials, Construction and Associated Component Systems; Their Decay and Treatment,* ed. J. Fidler, 19–28. English Heritage Research Transactions 2. London: James & James.

Finzi, E., V. Massa, and D. Morero. 1992. Radar-aided investigations on mosaics. In *Proceedings of the 7th International Congress on Deterioration and Conservation of Stone, Lisbon, 15–18 June 1992,* ed. J. Delgado Rodrigues, F. Henriques, and F. Telmo Jeremias, 969–78. Lisbon: Laboratório Nacional de Engenharia Civil.

Fitz, S. 1991. MONUFAKT: The Federal Environmental Agency's database for the protection of historic monuments and cultural heritage. In *Science, Technology, and European Cultural Heritage: Proceedings of the European Symposium, Bologna, Italy, 13–16 June 1989,* ed. N. S. Baer, C. Sabbioni, and A. I. Sors, 830–33. Oxford and Boston: Published for the Commission of the European Communities by Butterworth-Heinemann Publishers.

Fitz, S. 1996. MONUFAKT: A database system for improved information transfer in conservation. In *Proceedings of the 8th International Congress on Deterioration and Conservation of Stone, Berlin, 30 Sept.–4 Oct. 1996,* ed. J. Riederer, 1745–50. Berlin: Möller Druck und Verlag.

Fitzner, B. 2004. Documentation and evaluation of stone damage on monuments. In *Proceedings of the 10th International Congress on Deterioration and Conservation of Stone, Stockholm, June 27–July 2, 2004,* ed. D. Kwiatkowski and R. Löfvendahl, vol. 2, 677–90. Stockholm: ICOMOS Sweden.

Fitzner, B., and K. Heinrichs. 2002. Damage diagnosis on stone monuments: Weathering forms, damage categories and damage indices. In *Understanding and Managing of Stone Decay (SWAPNET 2001),* ed. R. Přikryl and H. A. Viles, 11–56. Prague: Charles University, The Karolinum Press.

Fitzner, B., K. Heinrichs, and R. Kownatzki. 1997. Weathering forms at natural stone monuments: Classification, mapping and evaluation. *Restoration of*

Buildings and Monuments: An International Journal = Bauinstandsetzen und Baudenkmalpflege: Eine internationale Zeitschrift 3 (2): 105–24.

Fitzner, B., K. Heinrichs, and D. La Bouchardiere. 2004. The Bangudae Petroglyph in Ulsan, Korea: Studies on weathering damage and risk prognosis. *Environmental Geology* 46 (3–4): 504–26.

Flatt, R. J. 2006. Hydration pressure: A bad case of crystallization pressure. In *Proceedings of the ARCCHIP Workshop ARIADNE 13: Problems of Salts in Masonry; "SALTeXPERT" (with the Getty Conservation Institute), November 27–December 1, 2002,* ed. S. Simon and M. Drdácký, vol. 5, 147–60. European Research on Cultural Heritage: State-of-the-Art Studies. Prague: Institute for Theoretical and Applied Mechanics, Academy of Sciences of the Czech Republic.

Flatt, R. J., and G. W. Scherer. 2002. Hydration and crystallization pressure of sodium sulfate: A critical review. In *Materials Issues in Art and Archaeology VI: Symposium Held Boston, November 26–30, 2001,* ed. P. B. Vandiver, M. Goodway, and J. L. Mass, 29–34. Materials Research Society Symposium Proceedings 712. Warrendale, PA: Materials Research Society.

Fontaine, L., M. L. Thomson, and G. T. Suter. 1999. Practice and research: The need for standards for historic mortars. In *The Use of and Need for Preservation Standards in Architectural Conservation,* ed. L. B. Sickles-Taves, 158–71. ASTM Special Technical Publication 1355. W. Conshohocken, PA: ASTM.

Ford, B., and K. Officer. 2005. Micro-environmental study of lichen invasion at a Ngunnawal Aboriginal rock art site in the eastern Australian sub-alpine region. In *14th Triennial Meeting, ICOM Committee for Conservation, The Hague, 12–16 September 2005: Preprints,* ed. I. Verger, 373–80. London: James & James, Earthscan.

Franke, L., J. Kiekbusch, R. Espinosa, and C. Gunstmann. 2007. CESA and ASTRA: Two program systems for cement and salt chemistry and the prediction of corrosion processes in concrete. In *Transport in Concrete: Nano- to Macrostructure: Proceedings of 5th International Essen Workshop, University of Duisburg-Essen, June 2007,* ed. M. J. Setzer, Freiburg: Aedificatio Verlag.

Franzen, C., and P. W. Mirwald. 2009. Moisture sorption behaviour of salt mixtures in porous stone. *Chemie der Erde—Geochemistry* 69 (1): 91–98.

Franzen, C., F. Hoferick, S. Laue, and H. Siedel. 2008. Water bath desalination of sandstone objects. In *Proceedings of the 11th International Congress on Deterioration and Conservation of Stone, 15–20 September 2008, Toruń, Poland,* ed. J. W. Lukaszewicz and P. Niemcewicz, 881–88. Toruń, Poland: Nicolaus Copernicus University.

Fratini, F., C. Manganelli del Fa, E. Pecchioni, G. Quarta, and A. Scala. 1990. Pietra di Lecce: Studio sull'applicazione di nuovi prodotti per la sua conservazione. In *Conservation of Monuments in the Mediterranean Basin: The Influence of Coastal Environment and Salt Spray on Limestone Marble; Proceedings of the 1st International Symposium, Bari, 7–10 June 1989 = La conservazione dei monumenti nel bacino del Mediterraneo: Influenza dell'ambiente costiero e dello spray marino sulla pietra calcarea e sul marmot; Atti del 1° simposio internazionale, Bari, 7–10 giugno 1989,* ed. F. Zezza, 495–99. Brescia, Italy: Grafo Edizioni.

Freie Universität, Berlin. 2006. Dahlem Workshops = Dahlem Konferenzen. http://www.fu-berlin.de/veranstaltungen/dahlemkonferenzen/en/index.html.

Friese, P., and A. Protz. 1997. Entsalzung mit zwei-seitiger Kompresse. *Bautenschutz + Bausanierung* 20 (8): 10–13.

Friese, P., A. Protz, and D. Peschl. 1997. Entsalzung: Kompressenputz mit passenden Porenradien. *Bautenschutz + Bausanierung* 20 (5): 26–30.

Furlan, V. 1992. Mesure du dépôt de soufre "sec" et rèactivité intrinseque despierres. In *Weathering and Air Pollution: Primo corso della Scuola universitaria C.U.M. conservazione dei monumenti, Lago di Garda (Portese), Venezia, Milano, 2–9 settembre 1991,* ed. F. Zezza, 131–38. Bari: Mario Adda Editore.

Gabrielli, N. 1991. Mural paintings. In *Science, Technology, and European Cultural Heritage: Proceedings of the European Symposium, Bologna, Italy, 13–16 June 1989,* ed. N. S. Baer, C. Sabbioni, and A. I. Sors, 168–79. Oxford and Boston: Published for the Commission of the European Communities by Butterworth-Heinemann Publishers.

Galán, E., and M. I. Carretero. 1994. Estimation of the efficacy of conservation treatments applied to a permotriassic sandstone. In *Conservation of Monuments in the Mediterranean Basin: Stone Monuments, Methodologies for the Analysis of Weathering and Conservation; Proceedings of the 3rd International Symposium, Venice, 22–25 June 1994 = La conservazione dei monumenti nel bacino del Mediterraneo: Materiali lapidei e monumenti, metodologie per l'analisi del degrado e la conservazione; Atti del 3° simposio internazionale, Venezia, 22–25 giugno 1994,* ed. V. Fassina, H. Ott, and F. Zezza, 947–54. Venice: Soprintendenza ai beni artistici e storici di Venezia.

Galán, E., M. Angeles Guerrero, M. Auxiliadora Vazquez, F. Maert, and F. Zezza. 1992. Marble weathering: Relation between ultrasonic data and macroscopic observations; The case of the columns of the Court of the Lions at the Alhambra in Granada, Spain. In *La conservation des monuments dans le bassin méditerranéen: Actes du 2ème Symposium international = The Conservation of Monuments in the Mediterranean Basin: Proceedings of the 2nd International Symposium,* ed. D. Decrouez, J. Chamay, and F. Zezza, 193–207. Geneva: Ville de Genève, Muséum d'histoire naturelle, Musée d'art et d'histoire.

Gardei, A., O. Garcia, M. Riedl, I. Vanhellemond, J. Strupi Suput, M.-L. Santarelli, I. Rodríguez-Maribona, and U. Müller. 2008. Performance and durability of a new antigraffiti system for cultural heritage: The EC project GRAFFITAGE. In *Proceedings of the 11th International Congress on Deterioration and Conservation of Stone, 15–20 September 2008, Toruń, Poland,* ed. J. W. Lukaszewicz and P. Niemcewicz, 889–97. Toruń, Poland: Nicolaus Copernicus University.

Gauri, K. L. 1974. Efficiency of epoxy resins as stone preservatives. *Studies in Conservation* 19 (2): 100–101.

Gauri, K. L., and M. V. Appa Rao. 1978. Certain epoxies, fluorocarbon-acrylics, and silicones as stone preservatives. In *Decay and Preservation of Stone: Prepared for the Engineering Geology Division of the Geological Society of America,* ed. E. M. Winkler and P. P. Hudec, 73–79. Engineering Geology Case Histories 11. Boulder, CO: The Geological Society of America.

Gauri, K. L., and J. K. Bandyopadhyay. 1999. *Carbonate Stone: Chemical Behavior, Durability, and Conservation.* New York: Wiley.

Gauri, K. L., and A. N. Chowdhury. 1988. Experimental studies on conversion of gypsum to calcite by microbes. In *Proceedings of the 6th International Congress on Deterioration and Conservation of Stone, Toruń, 12–14.09.1988,* ed. J. Ciabach, 545–50. Toruń, Poland: Nicolaus Copernicus University.

Gauri, K. L., L. Parks, J. Jaynes, and R. Atlas. 1992. Removal of sulphated-crust from marble using sulphate-reducing bacteria. In *Stone Cleaning and the Nature, Soiling and Decay Mechanisms of Stone: Proceedings of the International Conference Held in Edinburgh, UK, 14–16 April 1992,* ed. R. G. M. Webster, 160–65. London: Donhead Publishing.

Gaviño, M., B. Hermosin, M. Castillejo, M. Oujja, E. Rebollar, V. Vergès-Belmin, W. Nowik, and C. Saiz-Jimenez. 2004. Black crusts removal: The effect of stone yellowing and cleaning strategies. In *Air Pollution and Cultural Heritage: Proceedings of the International Workshop on Air Pollution and Cultural Heritage, 1–3 December 2003, Seville, Spain,* ed. C. Saiz-Jimenez, 239–46. Leiden and New York: Balkema.

Gaviño, M., B. Hermosin, V. Vergès-Belmin, W. Nowik, and C. Saiz-Jimenez. 2005. New insights on the chemical nature of stone yellowing produced after laser cleaning. In *Cultural Heritage Conservation and Environmental Impact Assessment by Non-Destructive Testing and Micro-Analysis,* ed. R. Van Grieken and K. Janssens, 149–58. Leiden: Balkema.

Gembinski, C., A. E. Charola, B. Price, and A. R. McGhie. 2000. Application of TGA/DTA/MS to the historic preservation of stone. *Journal of Thermal Analysis and Calorimetry* 59 (1): 601–9.

Getty Conservation Institute (GCI). 2009. Middle Eastern Geodatabase for Antiquities - Jordan (MEGA-J). Available at http://www.getty.edu/conservation/field_projects/jordan/.

Getty Conservation Institute (GCI) and Instituto Hondureño de Antropología e Historia (IHAH). 2006. Comparison of stairway photographic documentation through time. In *The Hieroglyphic Stairway of Copán, Honduras: Study Results and Conservation Proposals: A Project Report,* 27–33. Los Angeles: Getty Conservation Institute; Tegucigalpa: Instituto Hondureño de Antropología e

Historia. http://www.getty.edu/conservation/publications/pdf_publications/copan.html.

Giavarini, C., M. L. Santarelli, R. Natalini, and F. Freddi. 2008. A non-linear model of sulphation of porous stones: Numerical simulations and preliminary laboratory assessments. *Journal of Cultural Heritage* 9 (1): 14–22.

Gibson, L. T., B. G. Cooksey, D. Littlejohn, K. Linnow, M. Steiger, and N. H. Tennent. 2005. The mode of formation of thecotrichite, a widespread calcium acetate chloride nitrate efflorescence. *Studies in Conservation* 50 (4): 284–94.

Ginell, W. S., and R. Coffman. 1998. Epoxy resin-consolidated stone: Appearance change on aging. *Studies in Conservation* 43 (4): 242–48.

Ginell, W. S., D. Wessel, and C. Searles. 2001. *ASTM E2167 – 01 Standard Guide for Selection and Use of Stone Consolidants*. West Conshohocken, PA: ASTM International.

Giorgi, R., L. Dei, and P. Baglioni. 2000. A new method for consolidating wall paintings based on dispersions of lime in alcohol. *Studies in Conservation* 45 (3): 154–61.

Gnudi, C., R. Rossi-Manaresi, and O. Nonfarmale. 1979. *Notizie sul restauro della facciata di San Petronio = Report on the Conservation of the Facade of San Petronio*. Bologna: [Ministero per i beni culturali e ambientali, Centro "Cesare Gnudi" per la conservazione delle sculture all'aperto].

Gómez-Alarcon, G., and M. Angeles de la Torre. 1994. The effect of filamentous fungi on stone monuments: The Spanish experience. In *Building Mycology: Management of Decay and Health in Buildings,* ed. J. Singh, 295–309. London and New York: E & FN Spon.

Gómez-Heras, M., ed. 2007. Abstracts from SWAPNET 2007 and workshop on limestone decay and conservation (Malta, 24–26 May), Special issue, *Stone: Newsletter on Stone Decay* (2 August). http://www.qub.ac.uk/geomaterials/weathering/newsletter/issue_2.pdf.

Gómez-Heras, M., B. J. Smith, and R. Fort. 2006. Surface temperature differences between minerals in crystalline rocks: Implications for granular disaggregation of granites through thermal fatigue. *Geomorphology* 78 (3–4): 236–49.

Gómez-Heras, M., B. J. Smith, and R. Fort. 2008. Influence of surface heterogeneities of building granite on its thermal response and its potential for the generation of thermoclasty. *Environmental Geology* 56 (3–4): 547–60.

Gonçalves, T. D., and J. Delgado Rodrigues. 2006. Evaluating the salt content of salt-contaminated samples on the basis of their hygroscopic behavior, Part I: Fundamentals, scope and accuracy of the method. *Journal of Cultural Heritage* 7 (2): 79–84.

Gonçalves, T. D., J. Delgado Rodrigues, and M. M. Abreu. 2006. Evaluating the salt content of salt-contaminated samples on the basis of their hygroscopic behavior, Part II: Experiments with nine common soluble salts. *Journal of Cultural Heritage* 7 (3): 193–200.

Gonçalves, T. D., L. Pel, and J. Delgado Rodrigues. 2009. Influence of paints on drying and salt distribution processes in porous building materials. *Construction and Building Materials* 23 (5): 1751–59.

Gonçalves, L., C. C. Fonte, E. N. B. S. Júlio, and M. Caetano. 2009. Assessment of the state of conservation of buildings through roof mapping using very high spatial resolution images. *Construction and Building Materials* 23 (8): 2795–802.

Gonzales-del Valle, M., C. Dorronsoro, A. Irastorza, M. Duenas, S. Velasco, I. Ibarburu, and C. Saiz-Jimenez. 2003. Microbial communities in black crusts: An approach for assessing carbon utilisation. In *Molecular Biology and Cultural Heritage: Proceedings of the International Congress on Molecular Biology and Cultural Heritage, 4–7 March 2003, Sevilla, Spain,* ed. C. Saiz-Jimenez, 219–23. Lisse, Netherlands, and Exton, PA: Balkema.

Gorbushina, A. A. 2007. Life on the rocks. *Environmental Microbiology* 9 (7): 1613–31.

Gorbushina, A. A., E. Diakumaku, L. Mueller, and W. E. Krumbein. 2003. Biocide treatment of rock and mural paintings: Problems of application, molecular techniques of control and environmental hazards. In *Molecular Biology and Cultural Heritage: Proceedings of the International Congress on Molecular Biology and Cultural Heritage, 4–7 March 2003, Sevilla, Spain,* ed. C. Saiz-Jimenez, 61–72. Lisse, Netherlands, and Exton, PA: Balkema.

Gordon Research Conferences (GRC). 2010. Gordon Research Conferences. Available at http://www.grc.org/history.aspx.

Goudie, A. 1995. *The Changing Earth: Rates of Geomorphological Processes*. The Natural Environment. Oxford, UK, and Cambridge, MA: Blackwell.

Goudie, A. S., and H. A. Viles. 1999. The frequency and magnitude concept in relation to rock weathering. *Zeitschrift für Geomorphologie, Supplementband* 115: 175–89.

Goudie, A., and H. Viles. 2008. Weathering processes and forms. In *The History of the Study of Landforms or The Development of Geomorphology*, ed. T. P. Burt, R. J. Chorley, D. Brunsden, N. J. Cox, and A. S. Goudie, 129–64. London: Geological Society of London.

Gran, K. 2008. Deterioration of rock art in Scandinavia through frost thaw fluctuations and methods for reducing surface disintegration. In *Proceedings of the 11th International Congress on Deterioration and Conservation of Stone, 15–20 September 2008, Toruń, Poland*, ed. J. W. Lukaszewicz and P. Niemcewicz, 899–906. Toruń, Poland: Nicolaus Copernicus University.

Grassegger, G. 1999. Decay mechanisms of natural building stones on monuments: A review of the latest theories. In *Werkstoffe und Werkstoffprüfung im Bauwesen: Festschrift zum 60. Geburtstag von H.-W. Reinhardt*, ed. C. U. Große, 54–81. Stuttgart: IWB.

Grelk, B., C. Christiansen, B. Schouenborg, and K. Malaga. 2007. Durability of marble cladding: A comprehensive literature review. *Journal of ASTM International* 4 (4). DOI: 10.1520/JAI100857.

Grimmer, A. E., ed. 1984. *A Glossary of Historic Masonry Deterioration Problems and Preservation Treatments*. Washington, DC: Dept. of the Interior, National Park Service, Preservation Assistance Division (US Government Printing Office).

Grinzato, E., S. Marinetti, P. G. Bison, M. Concas, and S. Fais. 2004. Comparison of ultrasonic velocity and IR thermography for the characterisation of stones. Special issue, *Infrared Physics and Technology* 46 (1–2): 63–68.

Grissom, C. A., A. E. Charola, and M. J. Wachowiak. 2000. Measuring surface roughness on stone: Back to basics. *Studies in Conservation* 45 (2): 73–83.

Griswold, J., and S. Uricheck. 1998. Loss compensation methods for stone. *Journal of the American Institute for Conservation* 37 (1): 89–110.

Grossenbacher, Armin. 2009. To build an ecosystem of data on the Web. http://blogstats.wordpress.com/2009/07/20/to-build-an-ecosystem-of-data-on-the-web/.

Grossi, C. M., and M. Murray. 1999. Characteristics of carbonate building stones that influence the dry deposition of acidic gases. *Construction and Building Materials* 13 (3): 101–8.

Grossi, C. M., P. Brimblecombe, B. Menéndez, D. Benavente, and I. Harris. 2008. Long term change in salt weathering of stone monuments in north-west France. In *Proceedings of the 11th International Congress on Deterioration and Conservation of Stone, 15–20 September 2008, Toruń, Poland*, ed. J. W. Lukaszewicz and P. Niemcewicz, 121–28. Toruń, Poland: Nicolaus Copernicus University.

Grossi, C. M., R. M. Esbert, F. Díaz-Pache, and F. J. Alonso. 2003. Soiling of building stones in urban environments. *Building and Environment* 38 (1): 147–59.

Gruber, N., and J. N. Galloway. 2008. An earth-system perspective of the global nitrogen cycle. *Nature* 451 (7176): 293–96.

Guidetti, V., M. Chiavarini, and P. Parrini. 1992. Polyfluorourethanes as stone protectives. In *Proceedings of the 7th International Congress on Deterioration and Conservation of Stone, Lisbon, 15–18 June 1992*, ed. J. Delgado Rodrigues, F. Henriques, and F. Telmo Jeremias, 1279–88. Lisbon: Laboratório Nacional de Engenharia Civil.

Gulker, G., K. D. Hinsch, and A. El Jarad. 2004. Monitoring of salt-induced deformations in porous systems by microscopic speckle pattern interferometry. In *Proceedings of the 10th International Congress on Deterioration and Conservation of Stone, Stockholm, June 27–July 2, 2004*, ed. D. Kwiatkowski and R. Löfvendahl, vol. 2, 619–26. Stockholm: ICOMOS Sweden.

Gulotta, D., Cristina Tedeschi, Luigia Binda, and L. Toniolo. 2008. Durability of porous material to salt decay. In *Proceedings of the 11th International Congress on Deterioration and Conservation of Stone, 15–20 September 2008, Toruń,*

Poland, ed. J. W. Lukaszewicz and P. Niemcewicz, 129–36. Toruń, Poland: Nicolaus Copernicus University.

Haake, S., S. Simon, and M. Favaro. 2004. The Bologna cocktail: Evaluation of the consolidation treatments on monuments in France and Italy after 20 years of natural aging. In *Proceedings of the 10th International Congress on Deterioration and Conservation of Stone, Stockholm, June 27–July 2, 2004,* ed. D. Kwiatkowski and R. Löfvendahl, vol. 1, 423–30. Stockholm: ICOMOS Sweden.

Hall, C., and W. D. Hoff. 2007. Rising damp: Capillary rise dynamics in walls. *Proceedings of the Royal Society A: Mathematical, Physical and Engineering Sciences* 463 (2084): 1871–84.

Hall, C., W. D. Hoff, and A. Hamilton. 2008. Capillary rise dynamics in masonry structures: The role of water flux in long-term damage. In *Proceedings of the 11th International Congress on Deterioration and Conservation of Stone, 15–20 September 2008, Toruń, Poland,* ed. J. W. Lukaszewicz and P. Niemcewicz, 635–42. Toruń, Poland: Nicolaus Copernicus University.

Hall, K., and M.-F. André. 2006. Temperature observations in Antarctic tafoni: Implications for weathering, biological colonization, and tafoni formation. *Antarctic Science* 18 (3): 377–84.

Hall, K., I. Meiklejohn, and J. Arocena. 2007. The thermal responses of rock art pigments: Implications for rock art weathering in southern Africa. *Geomorphology* 91 (1–2): 132–45.

Halsey, D. P. 2000. Studying rock weathering with microcatchment experiments. *Zeitschrift für Geomorphologie, Supplementband* 120: 23–32.

Hamilton, A., and C. Hall. 2004. Mechanics of sodium sulphate crystal growth in the deterioration of sandstones: Some new observations. In *Proceedings of the 10th International Congress on Deterioration and Conservation of Stone, Stockholm, June 27–July 2, 2004,* ed. D. Kwiatkowski and R. Löfvendahl, vol. 1, 195–202. Stockholm: ICOMOS Sweden.

Hamilton, A., C. Hall, and L. Pel. 2008. Sodium sulfate heptahydrate: Direct observation of crystallization in a porous material. *Journal of Physics D: Applied Physics* 41 (21): 212002–7. DOI: 10.1088/0022-3727/41/21/212002.

Hamilton, A., V. Koutsos, and C. Hall. 2010. Direct measurement of salt-mineral repulsion using atomic force microscopy. *Chemical Communications* 46 (29): 5235–37.

Hammecker, C. 1993. Importance des transferts d'eau dans la dégradation des pierres en oeuvre. PhD diss., University of Strasbourg.

Hammecker, C., R. M. Esbert Alemany, and D. Jeannette. 1992. Geometry modifications of porous network in carbonate rocks by ethyl silicate treatment. In *Proceedings of the 7th International Congress on Deterioration and Conservation of Stone, Lisbon, 15–18 June 1992,* ed. J. Delgado Rodrigues, F. Henriques, and F. Telmo Jeremias, 1053–62. Lisbon: Laboratório Nacional de Engenharia Civil.

Hangal, S. P., and E. Harwit. 1997. Saving the Taj Mahal from air pollution. In *Proceedings of the Air and Waste Management Association's 90th Annual Meeting and Exhibition, 8–13 June 1997, Toronto.* Air & Waste Management Association TP6701-1 to TP6701/8. Pittsburgh: Air & Waste Management Association.

Hanna, S. B. 1984. The use of organo-silanes for the treatment of limestone in an advanced state of deterioration. In *Adhesives and Consolidants: IIC Preprints of the Contributions to the Paris Congress, 2–8 September 1984,* ed. N. S. Brommelle, E. M. Pye, P. Smith, and Garry Thomson, 171–76. London: International Institute for Conservation of Historic and Artistic Works.

Hansen, E. F., and N. Agnew. 1990. Consolidation with moisture curable isocyanates: Polyureas and polyurethanes. In *9th Triennial Meeting, ICOM Committee for Conservation, Dresden, German Democratic Republic, 26–31 August 1990: Preprints,* ed. K. Grimstad, 557–62. Los Angeles: ICOM Committee for Conservation; Marina del Rey, CA: Getty Conservation Institute.

Hansen, E., E. Doehne, J. Fidler, J. Larson, B. Martin, M. Matteini, C. Rodríguez-Navarro, E. Sebastián. Pardo, C. Price, A. de Tagle, J.-M. Teutonico, and N. R. Weiss. 2003. A review of selected inorganic consolidants and protective treatments for porous calcareous materials. *Reviews in Conservation* 4: 13–25.

Hauff, G., P. Kozub, and G. D'ham. 2008. Which cleaning method is the most appropriate one? A systematic approach to the assessment of cleaning test panels. In *Proceedings of the 11th International Congress on Deterioration and Conservation of Stone, 15–20 September 2008, Toruń, Poland,* ed. J. W. Lukaszewicz and P. Niemcewicz, 381–88. Toruń, Poland: Nicolaus Copernicus University.

Hawass, Z. A. 1998. *The Secrets of the Sphinx: Restoration Past and Present.* Cairo and New York: American University in Cairo Press.

Heinrichs, K. 2008. Diagnosis of weathering damage on rock-cut monuments in Petra, Jordan. *Environmental Geology* 56 (3–4): 643–75.

Heldal, T., E. Bloxam, and P. Storemyr. 2007. Unravelling ancient stone quarry landscapes in the Eastern Mediterranean: Three Egyptian case studies. In *Broadening Horizons: Multidisciplinary Approaches to Landscape Study,* ed. B. Ooghe and G. Verhoeven, 90–112. Newcastle, UK: Cambridge Scholars Publishing.

Heldal, T., E. Bloxam, P. Degryse, P. Storemyr, and A. Kelany. 2006. Old Kingdom gypsum quarries at Umm es-Sawan, Northern Faiyum Quarry Landscape, Egypt. In *Conservation of Ancient Stone Quarry Landscapes in the Eastern Mediterranean: First QuarryScapes Symposium, October 15–17, 2006, Antalya, Turkey: Extended Abstract Collection,* ed. P. Degryse, 11–14. Trondheim, Norway: QuarryScapes Project. http://www.quarryscapes.no/text/Workshops/Proceedings_1stQS_symposium.pdf#page=11.

Hempel, K. 1976. An improved method for the vacuum impregnation of stone. *Studies in Conservation* 21 (1): 40–43.

Henriques, F. M. A. 1992. Water repellents for masonry surfaces: A comparative analysis of performance tests. In *Proceedings of the 7th International Congress on Deterioration and Conservation of Stone: Held in Lisbon, Portugal, 15–18 June 1992,* ed. J. Delgado Rodrigues, Fernando Henriques, and F. Telmo Jeremias, 63140. Lisbon: Laboratório Nacional de Engenharia Civil.

Henry, A. 2006. *Stone Conservation: Principles and Practice.* Shaftesbury, UK: Donhead Publishing.

Heritage, A., A. Heritage, F. Funke, V. Vergès-Belmin, and A. Bourgès. 2008. How do conservators tackle desalination? An international survey of current poulticing methods. In *Cultural Heritage Research Meets Practice: 8th European Conference on Research for Protection, Conservation and Enhancement of Cultural Heritage, November 10–12, 2008, Ljubljana, Slovenia; Conference Preprints,* ed. J. Kolar and M. Strli, 58–61. Ljubljana, Slovenia: National and University Library. http://www.si.edu/mci/downloads/CHRESP%202008/CHRESP_proof_2008_LJUBLJANA.pdf.

Hermosin, B., M. Gaviño, and C. Saiz-Jimenez. 2004. Organic compounds in black crusts from different European monuments: A comparative study. In *Air Pollution and Cultural Heritage: Proceedings of the International Workshop on Air Pollution and Cultural Heritage, 1–3 December 2003, Seville, Spain,* ed. C. Saiz-Jimenez, 47–56. Leiden and New York: Balkema.

Hernanz, A., J. M. Gavira-Vallejo, J. F. Ruiz-López, and H. G. M. Edwards. 2008. A comprehensive micro-Raman spectroscopic study of prehistoric rock paintings from the Sierra de las Cuerdas, Cuenca, Spain. Special issue, *Journal of Raman Spectroscopy* 39 (8): 972–84.

Herrera, L. K., S. Le Borgne, and H. A. Videla. 2009. Modern methods for materials characterization and surface analysis to study the effects of biodeterioration and weathering on buildings of cultural heritage. *International Journal of Architectural Heritage* 3 (1): 74–91.

Herrmann, J., N. Herz, and R. Newman, eds. 2002. *ASMOSIA V: Interdisciplinary Studies on Ancient Stone; Proceedings of the Fifth International Conference of the Association for the Study of Marble and Other Stones in Antiquity, Museum of Fine Arts, Boston, 1998.* ASMOSIA 5. London: Archetype.

Hoerlé, S. 2005. A preliminary study of the weathering activity at the rock art site of Game Pass Shelter (KwaZulu-Natal, South Africa) in relation to its conservation. *South African Journal of Geology* 108 (2): 297–308.

Hoerlé, S. 2006. Rock temperatures as an indicator of weathering processes affecting rock art. *Earth Surface Processes and Landforms* 31 (3): 383–89.

Hoffmann, M., and H. Heuser. 1993. The injector jet process for the cleaning of natural stone facades. In *Conservation of Stone and Other Materials: Proceedings of the International RILEM/UNESCO Congress "Conservation of Stone and Other Materials: Research – Industry – Media," UNESCO Headquarters, Paris, June 29–July 1, 1993,* ed. M.-J. Thiel, 526–33. RILEM Proceedings 21. London and New York: E & FN Spon.

Holm, A., and H. M. Künzel. 2003. Two-dimensional transient heat and moisture simulations of rising damp with WUFI 2d. In *Research in Building Physics: Proceedings of the 2nd International Conference on Building Physics, 14–18 September 2003, Antwerp, Belgium,* ed. J. Carmeliet, H. S. L. C. Hens, and G. Vermeir, 363–67. Lisse, Netherlands, and Exton, PA: Balkema. http://www.hoki.ibp.fhg.de/ibp/publikationen/konferenzbeitraege/pub1_6.pdf.

Holmer, B. 1998. Flaking by insolation drying and salt weathering on the Swedish west coast. *Zeitschrift für Geomorphologie* 42 (1): 39–55.

Holmes, L. L., G. Harbottle, and A. Blanc. 1994. Compositional characterization of French limestone: A new tool for art historians. *Archaeometry* 36 (1): 25–39.

Honeyborne, D. B. 1998. Weathering and decay of masonry. In *Conservation of Building and Decorative Stone,* ed. J. Ashurst and F. G. Dimes, 153–78. Butterworth-Heinemann Series in Conservation and Museology. Oxford and Woburn, MA: Butterworth-Heinemann.

Honeyborne, D., J. Ashurst, C. Price, and K. Ross. 1990. Surface treatments. In *Conservation of Building and Decorative Stone,* ed. J. Ashurst and F. G. Dimes, vol. 2, 155–84. Butterworth-Heinemann Series in Conservation and Museology. London and Boston: Butterworth-Heinemann.

House of Lords, Science and Technology Committee. 2006. *Science and Heritage: Report with Evidence.* HL Paper 256. London: Stationery Office. http://www.publications.parliament.uk/pa/ld200506/ldselect/ldsctech/256/256.pdf.

House of Lords, Science and Technology Committee. 2007. *Science and Heritage: An Update Report with Evidence.* HL Paper 168. London: Stationery Office. http://www.publications.parliament.uk/pa/ld200607/ldselect/ldsctech/168/168.pdf.

Howe, E. 2007. *A Report on the Conservation of the Lichfield Angel.* Birmingham: Birmingham Museum & Galleries Collection Centre. http://lichfield-angel.net/data/LichfieldAngel_Conservation_Howe.pdf.

Hoyos, M., J. C. Cañaveras, S. Sánchez-Moral, E. Sanz-Rubio, and V. Soler. 1998. Microclimatic characterization of a karstic cave: Human impact on microenvironmental parameters of a prehistoric rock art cave (Candamo Cave, northern Spain). *Environmental Geology* 33 (4): 231–42.

Huinink, H. P., L. Pel, and K. Kopinga. 2004. Simulating the growth of tafoni. *Earth Surface Processes and Landforms* 29 (10): 1225–33.

Huinink, H. P., J. Petkovic, L. Pel, and K. Kopinga. 2006. Water and salt transport in plaster/substrate systems. *Heron* 51 (1): 9–31.

Huneau, F., S. Hoerlé, A. Denis, and A. Salomon. 2008. First use of geological radar to assess the conservation condition of a South African rock art site: Game Pass Shelter (KwaZulu-Natal). *South African Journal of Science* 104 (7–8): 251–54.

Hussein, A. S., and H. El-Shishiny. 2009. Influences of wind flow over heritage sites: A case study of the wind environment over the Giza Plateau in Egypt. *Environmental Modelling and Software* 24 (3): 389–410.

Hygen, A.-S. 2006. *Protection of Rock Art: The Rock Art Project 1996–2005; Final Report from the Directorate for Cultural Heritage.* Oslo: Directorate for Cultural Heritage Archive. http://www.riksantikvaren.no/filestore/rockart project-finalreport.pdf.

Hyslop, E. K. 2008. Assessing stone decay and the future repair needs of a sandstone city: A survey of the stone built heritage in Glasgow, Scotland. In *Proceedings of the 11th International Congress on Deterioration and Conservation of Stone, 15–20 September 2008, Toruń, Poland,* ed. J. W. Lukaszewicz and P. Niemcewicz, 117–24. Toruń, Poland: Nicolaus Copernicus University.

Hyslop, E., and A. McMillan. 2004. Replacement sandstone in the Edinburgh World Heritage Site: Problems of source and supply. In *Proceedings of the 10th International Congress on Deterioration and Conservation of Stone, Stockholm, June 27–July 2, 2004,* ed. D. Kwiatkowski and R. Löfvendahl, vol. 2, 777–84. Stockholm: ICOMOS Sweden.

Hyslop, E., A. McMillan, D. Cameron, A. Leslie, and G. Lott. 2009. Building stone databases in the UK: A practical resource for conservation. *Engineering Geology.* DOI: 10.1016/j.enggeo.2009.05.008.

Iglesias, M., J. L. Prada, and N. Guasch. 2008. Estudio de la técnica de limpieza de la caliza limosa dolomitizada del Mioceno de Tarragona alterada en ambiente urbano [Technique for cleaning Tarragona Miocene age dolomitized silty limestone, altered by urban pollution]. *Materiales de construccion* 58 (289–90): 247–62.

Iñigo, A. C., S. Vicente-Tavera, M. A. Vicente, and V. Rives. 2001. Effect of treatments on pore network of granites. *Journal of Materials in Civil Engineering* 13 (3): 166–68.

Inkpen, R. J., P. Collier, and D. Fontana. 2000. Close-range photogrammetric analysis of rock surfaces. *Zeitschrift für Geomorphologie, Supplementband* 120: 67–81.

Inkpen, R., B. Duane, P. Collier, J. Burdett, and T. Yates. 2004. Historic and contemporary patterns of weathering in Bath: Assessing weathering using an integrated database and geographical information system. In *Proceedings of the 10th International Congress on Deterioration and Conservation of Stone, Stockholm, June 27–July 2, 2004,* ed. D. Kwiatkowski and R. Löfvendahl, vol. 2, 667–76. Stockholm: ICOMOS Sweden.

Jackson, M. D., F. Marra, R. L. Hay, C. Cawood, and E. M. Winkler. 2005. The judicious selection and preservation of tuff and travertine building stone in ancient Rome. *Archaeometry* 47 (3): 485–510.

Jacobs, J. 2004. Interior cleaning of St Paul's: Wren uncovered. *Natural Stone Specialist* 39 (5): 22–24, 26, 28.

Jacobs, P., E. Sevens, and M. Kunnen. 1995. Principles of computerised X-ray tomography and applications to building materials. *Science of the Total Environment* 167: 161–70.

Jacobson, A. D., and L. Wu. 2009. Microbial dissolution of calcite at T = 28°C and ambient pCO2. *Geochimica et Cosmochimica Acta* 73 (8): 2314–31.

Jain, K. K., A. K. Mishra, and T. Singh. 1994. Biodeterioration of stone: A review of mechanisms involved. In *Recent Advances in Biodeterioration and Biodegradation,* ed. K. L. Garg, N. Garg, and K. G. Mukerji, vol. 1, 323–54. Calcutta: Naya Prokash.

Jaynes, S. M., and R. U. Cooke. 1987. Stone weathering in Southeast England. *Atmospheric Environment* 21 (7): 1601–22.

Jefferson, D., S. Hanna, B. Martin, and D. M. Jones. 2006. *Identifying and Sourcing Stone for Historic Building Repair: An Approach to Determining and Obtaining Compatible Replacement Stone.* Technical Advice Note (English Heritage). Swindon, UK: English Heritage. http://www.shropshiregeology.org.uk/building_stones/English_Heritage_Identifying_and_Sourcing_Stone.pdf.

Jeyaraj, V. 2004. Rock and cave paintings of the ancient past in India and their care. In *Studies in Art and Archaeological Conservation: Dr. B. B. Lal Commemoration Volume,* ed. A. S. Bisht and S. P. Singh, 91–100. Delhi: Agam Kala Prakashan.

Jie Chen, H. P. Blume, and L. Beyer. 2000. Weathering of rocks induced by lichen colonization–A review. *Catena* 39 (2): 121–46.

Jiménez-González, I., and G. W. Scherer. 2004. Effect of swelling inhibitors on the swelling and stress relaxation of clay bearing stones. *Environmental Geology* 46 (3–4): 364–77.

Jiménez-González, I., C. Rodríguez-Navarro, and G. W. Scherer. 2008. Role of clay minerals in the physicomechanical deterioration of sandstone. *Journal of Geophysical Research F: Earth Surface* 113 (2): F02021.

Jiménez-Lopez, C., F. Jroundi, C. Pascolini, C. Rodríguez-Navarro, G. Piñar-Larrubia, M. Rodríguez-Gallego, and M. T. González-Muñoz. 2008. Consolidation of quarry calcarenite by calcium carbonate precipitation induced by bacteria activated among the microbiota inhabiting the stone. *International Biodeterioration and Biodegradation* 62 (4): 352–63.

Jiménez-Lopez, C., C. Rodríguez-Navarro, G. Piñar, F. J. Carrillo-Rosúa, M. Rodríguez-Gallego, and M. T. González-Muñoz. 2007. Consolidation of degraded ornamental porous limestone stone by calcium carbonate precipitation induced by the microbiota inhabiting the stone. *Chemosphere* 68 (10): 1929–36.

Jones, M. S., and R. D. Wakefield. 1999. *Aspects of Stone Weathering, Decay and Conservation: Proceedings of the 1997 Stone Weathering and Atmospheric Pollution Network Conference (SWAPNET '97), The Robert Gordon University, 15–17 May 1997.* London: Imperial College Press.

Jurado, V., A. Fernandez-Cortes, S. Cuezva, L. Laiz, J. C. Cañaveras, S. Sanchez-Moral, and C. Saiz-Jimenez. 2009. The fungal colonisation of rock-art caves: Experimental evidence. *Naturwissenschaften* 96 (9): 1027–34.

Kaminski, K. 2008. Brick walls structural moisture content: A new approach to diagnostic procedure. In *Proceedings of the 11th International Congress on Deterioration and Conservation of Stone, 15–20 September 2008, Toruń, Poland,* ed. J. W. Lukaszewicz and P. Niemcewicz, 389–96. Toruń, Poland: Nicolaus Copernicus University.

Kapsalas, P., P. Maravelaki-Kalaitzaki, M. Zervakis, E. T. Delegou, and A. Moropoulou. 2007. Optical inspection for quantification of decay on stone surfaces. *NDT & E International* 40 (1): 2–11.

Karatasios, I., V. Kilikoglou, B. Colston, P. Theoulakis, and D. Watt. 2007. Setting process of lime-based conservation mortars with barium hydroxide. *Cement and Concrete Research* 37 (6): 886–93.

Karatasios, I., P. Theoulakis, A. Kalagri, A. Sapalidis, and V. Kilikoglou. 2009. Evaluation of consolidation treatments of marly limestones used in archaeological monuments. *Construction and Building Materials* 23 (8): 2803–12.

Keene, L., and Fu-Pen Chiang. 2009. Real-time anti-node visualization of vibrating distributed systems in noisy environments using defocused laser speckle contrast analysis. *Journal of Sound and Vibration* 320 (3): 472–81.

Keyser, J. D., M. Greer, and J. Greer. 2005. Arminto petroglyphs: Rock art damage assessment and management considerations in Central Wyoming. *Plains Anthropologist* 50 (193): 23–30.

Khummalai, N., and V. Boonamnuayvitaya. 2005. Suppression of arsenopyrite surface oxidation by sol-gel coatings. *Journal of Bioscience and Bioengineering* 99 (3): 277–84.

Kim, Eun Kyung, Jongok Won, Jeong-Jin Kim, Yong Soo Kang, and Sa Dug Kim. 2008. TEOS/GPTMS/silica nanoparticle solutions for conservation of Korean heritage stones. In *Proceedings of the 11th International Congress on Deterioration and Conservation of Stone, 15–20 September 2008, Toruń, Poland,* ed. J. W. Lukaszewicz and P. Niemcewicz, 915–23. Toruń, Poland: Nicolaus Copernicus University.

Kirchner, G. 1996. Cavernous weathering in the Basin and Range area, southwestern USA and northwestern Mexico. *Zeitschrift für Geomorphologie, Supplementband* 106: 73–97.

Kirkitsos, P., and D. Sikiotis. 1995. Deterioration of Pentelic marble, Portland limestone and Baumberger sandstone in laboratory exposures to gaseous nitric acid. *Atmospheric Environment* 29 (1): 77–86.

Klamma, R., M. Spaniol, M. Jarke, Yiwei Cao, M. Jansen, and G. Toubekis. 2006. A hypermedia Afghan sites and monuments database. In *Geographic Hypermedia: Concepts and Systems; Proceedings of the First International Workshop on Geographic Hypermedia, Denver, April 4–5, 2005,* ed. E. Stefanakis, 59–73. Lecture Notes in Geoinformation and Cartography. Berlin and New York: Springer. http://dbis.rwth-aachen.de/lehrstuhl/staff/klamma/download/Klamma-GeoHM05-crc.pdf.

Kleiner, Th., and Aloysius Wehr. 1994. Fast digital survey of historical sites and monuments by using the 4D-laser-scanner system. In *Optical 3D Measurement Techniques II: Applications in Inspection, Quality Control, and Robotics, 4–7 October 1993, Zürich,* ed. A. Gruen and H. Kahmen, 301–8. Proceedings of SPIE (The International Society for Optical Engineering) 2252. Bellingham, WA: SPIE.

Klemm, R., and D. D. Klemm. 2008. *Stone and Stone Quarries in Ancient Egypt.* Illustrated ed. London: British Museum Press.

Kloor, K. 2008. Archaeology: Dust storm rising over threat to famed rock art in Utah. *Science* 319 (5862): 394.

Koestler, R. J., and O. Salvadori. 1996. Methods of evaluating biocides for the conservation of porous building materials. *Science and Technology for Cultural Heritage* 5 (1): 63–68.

Koestler, R. J., and E. D. Santoro. 1988. *Assessment of the Susceptibility to Biodeterioration of Selected Polymers and Resins: Final Report Submitted to the Getty Conservation Institute.* GCI Scientific Program Reports. Bloomfield, NJ:

Bloomfield College; Marina del Rey, CA: Getty Conservation Institute, Scientific Program. http://www.getty.edu/conservation/publications/pdf_publications/assessment.pdf.

Konkol, N., C. McNamara, J. Sembrat, M. Rabinowitz, and R. Mitchell. 2009. Enzymatic decolorization of bacterial pigments from culturally significant marble. *Journal of Cultural Heritage* 10 (3): 362–66.

Koss, A., and J. Marczak. 2008. Implementation of laser technology in conservation: Last decade. In *Proceedings of the 11th International Congress on Deterioration and Conservation of Stone, 15–20 September 2008, Toruń, Poland,* ed. J. W. Lukaszewicz and P. Niemcewicz, 939–48. Toruń, Poland: Nicolaus Copernicus University.

Kouzeli, K. 1992. Black crust removal methods in use: Their effects on Pentelic marble surfaces. In *Proceedings of the 7th International Congress on Deterioration and Conservation of Stone, Lisbon, 15–18 June 1992,* ed. J. Delgado Rodrigues, F. Henriques, and F. Telmo Jeremias, 1147–56. Lisbon: Laboratório Nacional de Engenharia Civil.

Kozlowski, R., M. Tokarz, and M. Persson. 1992. "GYPSTOP": A novel protective treatment. In *Proceedings of the 7th International Congress on Deterioration and Conservation of Stone, Lisbon, 15–18 June 1992,* ed. J. Delgado Rodrigues, F. Henriques, and F. Telmo Jeremias, 1187–96. Lisbon: Laboratório Nacional de Engenharia Civil.

Kozub, P. 2004. New injection masses based on water-emulgated epoxide resins used to fill crevices in rose granite from Aswan in architectural elements of the temple of an Egyptian goddess Bastet from Tell Basta (Egypt). In *Proceedings of the 10th International Congress on Deterioration and Conservation of Stone, Stockholm, June 27–July 2, 2004,* ed. D. Kwiatkowski and R. Löfvendahl, vol. 1, 307–14. Stockholm: ICOMOS Sweden.

Krumbein, W. E., S. E. Diakumaku, K. Petersen, T. Warscheid, and C. Urzi. 1993. Interactions of microbes with consolidants and biocides used in the conservation of rocks and mural paintings. In *Conservation of Stone and Other Materials: Proceedings of the International RILEM/UNESCO Congress "Conservation of Stone and Other Materials: Research—Industry—Media," UNESCO Headquarters, Paris, June 29–July 1, 1993,* ed. M.-J. Thiel, 589–96. RILEM Proceedings 21. London and New York: E & FN Spon.

Krumbein, W. E., and C. Urzi. 1992. Biologically induced decay phenomena of antique marbles: Some general considerations. In *La conservation des monuments dans le bassin méditerranéen: Actes du 2ème Symposium international = The Conservation of Monuments in the Mediterranean Basin: Proceedings of the 2nd International Symposium,* ed. D. Decrouez, J. Chamay, and F. Zezza, 219–35. Geneva: Ville de Genève, Muséum d'histoire naturelle, and Musée d'art et d'histoire.

Kühlenthal, M., E. Kaiser, and H. Fischer. 2000. Testergebnisse und Verfahrensweisen bei der Einstellung des Reparaturmörtels. In *Petra: Die Restaurierung der Grabfassaden; Deutsch-Jordanisches Projekt zum Aufbau eines Konservierungszentrums für Petra = Petra: The Restoration of the Rockcut Tomb Façades: German-Jordanian Project for the Establishment of a Conservation Center in Petra,* ed. M. Kühlenthal, H. Fischer, and T. S. Akasheh, 201–8. ICOMOS-Hefte des Deutschen Nationalkomitees 34/Arbeitshefte des Bayerischen Landesamtes für Denkmalpflege 105. Munich: Bayerisches Landesamt für Denkmalpflege.

Kumar, R., and W. S. Ginell. 1995. Evaluation of consolidants for stabilization of weak Maya limestone. In *Methods of Evaluating Products for the Conservation of Porous Building Materials in Monuments: International Colloquium, Rome, 19–21 June 1995; Preprints,* 163–78. Rome: ICCROM.

Kumar, R., and C. A. Price. 1994. The influence of salts on the hydrolysis and condensation of methyl-trimethoxysilane. In *Conservation of Monuments in the Mediterranean Basin: Stone Monuments, Methodologies for the Analysis of Weathering and Conservation; Proceedings of the 3rd International Symposium, Venice, 22–25 June 1994 = La conservazione dei monumenti nel bacino del Mediterraneo: Materiali lapidei e monumenti, metodologie per l'analisi del degrado e la conservazione; Atti del 3° simposio internazionale, Venezia, 22–25 giugno 1994,* ed. V. Fassina, H. Ott, and F. Zezza, 861–65. Venice: Soprintendenza ai beni artistici e storici di Venezia.

Kumar, N., U. C. Kulshreshta, A. Saxena, K. M. Kumari, and S. S. Srivastava. 1993. Effect of anthropogenic activity on formate and acetate levels in precipitation at four sites in Agra, India. *Atmospheric Environment – Part B Urban Atmosphere* 27 B (1): 87–91.

Kun Hong and Yuzhong Zhan. 2008. Exploring research of biomimetic SiO_2 membranes on surfaces of historic stones. *Fushi yu fanghu = Corrosion & Protection* 29 (9): 520–22.

Lacanette, D., S. Vincent, A. Sarthou, P. Malaurent, and J.-P. Caltagirone. 2009. An Eulerian/Lagrangian method for the numerical simulation of incompressible convection flows interacting with complex obstacles: Application to the natural convection in the Lascaux cave. *International Journal of Heat and Mass Transfer* 52 (11–12): 2528–42.

Laiz, L., D. Recio, B. Hermosin, and C. Saiz-Jimenez. 2000. Microbial communities in salt efflorescences. In *Of Microbes and Art: The Role of Microbial Communities in the Degradation and Protection of Cultural Heritage,* ed. O. Ciferri, P. Tiano, and G. Mastromei, 77–88. New York: Kluwer Academic/Plenum Publishers.

Lambert, D. 1988. Graffiti removal on rock art sites, NSW. *AICCM Bulletin* 14 (1–2): 9–12. http://www.aiccm.org.au/docs/Bulletin1988/Lambert_Bulletin_1988_Vol14No1and2.pdf.

Larkin, M. 2002. Preserving the legacy of rock art. *Lancet* 359 (9318): 1707.

Larson, J. H. 1992. New approaches to the conservation of external stone sculpture: The twelfth century frieze at Lincoln Cathedral. In *Proceedings of the 7th International Congress on Deterioration and Conservation of Stone, Lisbon, 15–18 June 1992,* ed. J. Delgado Rodrigues, F. Henriques, and F. Telmo Jeremias, 1167–75. Lisbon: Laboratório Nacional de Engenharia Civil.

Larssen, T., E. Lydersen, Dagang Tang, Yi He, Jixi Gao, Haiying Liu, Lei Duan, H. M. Seip, R. D. Vogt, J. Mulder, Min Shao, Yanhui Wang, He Shang, Xiaoshan Zhang, S. Solberg, W. Aas, T. Okland, O. Eilertsen, V. Angell, Quanru Li, Dawei Zhao, Renjun Xiang, Jinshong Xiao, and Jiahai Luo. 2006. Acid rain in China. *Environmental Science and Technology* 40 (2): 418–25.

Laue, S., C. Bläuer Böhm, and D. Jeannette. 1996. Salt weathering and porosity: Examples from the crypt of St. Maria im Kapitol, Cologne. In *Proceedings of the 8th International Congress on Deterioration and Conservation of Stone, Berlin, 30 Sept.–4 Oct. 1996,* ed. J. Riederer, 513–22. Berlin: Möller Druck und Verlag.

Laue, S., H. Siedel, G. Hilbert, and S. Pfefferkorn. 2005. Alveolarverwitterung und schichtparalleles Auswittern der Kreidesandsteine an der Kirche von Leuba/Sachsen. *Zeitschrift der Deutschen Gesellschaft für Geowissenschaften* 156 (1): 141–49.

Laurenzi Tabasso, M. 1995. Acrylic polymers for the conservation of stone: Advantages and drawbacks. *APT Bulletin* 26 (4): 17–21.

Laurenzi Tabasso, M. 2008. Testing consolidants: A case study on the selection of consolidants for the plaster of wall paintings. In *Stone Consolidation in Cultural Heritage: Research and Practice; Proceedings of the International Symposium, Lisbon, 6–7 May 2008,* ed. J. Delgado Rodrigues and J. M. Mimoso, 15–28. Lisbon: LNEC (Laboratório Nacional de Engenharia Civil).

Laurenzi Tabasso, M., and S. Simon. 2006. Testing methods and criteria for the selection/evaluation of products for the conservation of porous building materials. *Reviews in Conservation* 7: 67–82.

Laver, M. E., and I. N. M. Wainwright. 1995. An investigation of the dissolution of a marble petroglyph site by acidic precipitation. *Studies in Conservation* 40 (4): 265–73.

Lazzarini, L., ed. 2002. *ASMOSIA VI: Interdisciplinary Studies on Ancient Stone; Proceedings of the Sixth International Conference of the Association for the Study of Marble and Other Stones in Antiquity, Venice, June 15–18, 2000.* ASMOSIA 6. Padua: Bottega d'Erasmo.

Lazzarini, L., and O. Salvadori. 1989. A reassessment of the formation of the patina called *scialbatura. Studies in Conservation* 34 (1): 20–26.

Lazzarini, L., F. Antonelli, S. Cancelliere, and A. Conventi. 2008. The deterioration of Euganean trachyte in Venice. In *Proceedings of the 11th International Congress on Deterioration and Conservation of Stone, 15–20 September 2008, Toruń,*

Poland, ed. J. W. Lukaszewicz and P. Niemcewicz, 153–63. Toruń, Poland: Nicolaus Copernicus University.

Le Métayer-Levrel, G., S. Castanier, G. Orial, J.-F. Loubière, and J.-P. Perthuisot. 1999. Applications of bacterial carbonatogenesis to the protection and regeneration of limestones in buildings and historic patrimony. *Sedimentary Geology* 126 (1–4): 25–34.

Leisen, H. 2002. Contour scaling: The disfiguring disease of Angkor Wat reliefs. *Museum International* 54 (1–2): 85–92.

Leisen, H., E. von Plehwe-Leisen, and S. Warrack. 2004. Success and limits for stone repair mortars based on tetra ethyl silicate: Conservation of the reliefs at Angkor Wat temple, Cambodia. In *Proceedings of the 10th International Congress on Deterioration and Conservation of Stone, Stockholm, June 27–July 2, 2004*, ed. D. Kwiatkowski and R. Löfvendahl, vol. 1, 331–38. Stockholm: ICOMOS Sweden.

Leitner, H. 2005. The treatment of wall paintings affected by salts: An interdisciplinary task as seen from a conservator's perspective. *Restoration of Buildings and Monuments: An International Journal = Bauinstandsetzen und Baudenkmalpflege: Eine Internationale Zeitschrift* 11 (6): 365.

LERM (Laboratoire d'Etudes et de Recherches sur les Matériaux). 2006. Le LERM Dossier technique: Elaboration de la norme pierre B 10-601: Le cas de la gélivité des pierres. *Newsletter* (3). http://www.lerm.fr/lerm/Newsletter/Newsletter3/lerm_Newsletter3_dossier_B10-601.shtml.

Leroux, L., V. Vergès-Belmin, D. Costa, J. Delgado Rodrigues, P. Tiano, R. Snethlage, B. Singer, S. Massey, and E. De Witte. 2000. Measuring the penetration depth of consolidating products: Comparison of six methods. In *Proceedings of the 9th International Congress on Deterioration and Conservation of Stone, Venice, June 19–24, 2000*, ed. V. Fassina, 361–69. Amsterdam and New York: Elsevier.

Lewin, S. Z. 1982. The mechanism of masonry decay through crystallization. In *Conservation of Historic Stone Buildings and Monuments: Report of the Committee on Conservation of Historic Stone Buildings and Monuments, National Materials Advisory Board, Commission on Engineering and Technical Systems, National Research Council*, 120–44. Washington, DC: National Academy Press.

Lewin, S. Z. 1988. The current state of the art in the use of synthetic materials for stone conservation. In *The Deterioration and Conservation of Stone: Notes from the International Venetian Courses on Stone Restoration*, ed. L. Lazzarini and R. Pieper. Studies and Documents on the Cultural Heritage 16. Paris: UNESCO.

Lewin, S. Z., and N. S. Baer. 1974. Rationale of the barium hydroxide-urea treatment of decayed stone. *Studies in Conservation* 19 (1): 24–35.

Lewis, F. J., E. May, and A. F. Bravery. 1988. Metabolic activities of bacteria isolated from building stone and their relationship to stone decay. In *Biodeterioration 7: Selected Papers Presented at the Seventh International Biodeterioration Symposium, Cambridge, UK, 6–11 September 1987*, ed. D. R. Houghton, R. N. Smith, and H. O. W. Eggins, 107–12. London and New York: Elsevier.

Licchelli, M., and S. J. Marzolla. 2008. Enhanced protection against mural writings by crosslinked fluorinated polymers. In *Proceedings of the 11th International Congress on Deterioration and Conservation of Stone, 15–20 September 2008, Toruń, Poland*, ed. J. W. Lukaszewicz and P. Niemcewicz, 967–74. Toruń, Poland: Nicolaus Copernicus University.

Lipfert, F. W. 1989. Atmospheric damage to calcareous stones: Comparison and reconciliation of recent experimental findings. *Atmospheric Environment* 23 (2): 415–29.

Lithgow, K., T. Curteis, and L. Bullock. 2007. Managing external environments through preventive conservation: The investigation and control of environmentally caused deterioration of decorative surfaces in the Marlborough Pavilion, Chartwell, Kent. In *Museum Microclimates: Contributions to the Copenhagen Conference, 19–23 November 2007*, ed. T. Padfield and K. Borchersen. Copenhagen: National Museum of Denmark. http://www.natmus.dk/graphics/konferencer_mm/microclimates/pdf/lithgow.pdf.

Littmann, K., H. R. Sasse, S. Wagener, and H. Höcker. 1993. Development of polymers for the consolidation of natural stone. In *Conservation of Stone and Other Materials: Proceedings of the International RILEM/UNESCO Congress*

"Conservation of Stone and Other Materials: Research—Industry—Media," UNESCO Headquarters, Paris, June 29–July 1, 1993, ed. M.-J. Thiel, 681–88. RILEM Proceedings 21. London and New York: E & FN Spon.

Livingston, R. A. 1992. Graphical methods for examining the effects of acid rain and sulfur dioxide on carbonate stones. In *Proceedings of the 7th International Congress on Deterioration and Conservation of Stone, Lisbon, 15–18 June 1992,* ed. J. Delgado Rodrigues, F. Henriques, and F. Telmo Jeremias, 375–86. Lisbon: Laboratório Nacional de Engenharia Civil.

Livingston, R. A. 1997. Development of air pollution damage functions. In *Saving Our Architectural Heritage: The Conservation of Historic Stone Structures; Report of the Dahlem Workshop on Saving Our Architectural Heritage, The Conservation of Historic Stone Structures, Berlin, March 3–8, 1996*, ed. N. S. Baer and R. Snethlage, 37–62. Dahlem Workshop Reports. Chichester and New York: John Wiley & Sons.

Livingston, R. A. 2008. Mass balance study of acid rain attack on the statue of Phoenicia. In *Proceedings of the 11th International Congress on Deterioration and Conservation of Stone, 15–20 September 2008, Toruń, Poland,* ed. J. W. Lukaszewicz and P. Niemcewicz, 165–72. Toruń, Poland: Nicolaus Copernicus University.

Lockwood, S. 1994. Reigate stone: Geology, use and repair. *Structural Survey* 12 (5): 18–22.

Lombardo, T., E. Doehne, and S. Simon. 2004. The response of NaCl and Umm Ishrin sandstone to humidity cycling: Mechanisms of salt weathering. In *Proceedings of the 10th International Congress on Deterioration and Conservation of Stone, Stockholm, June 27–July 2, 2004,* ed. D. Kwiatkowski and R. Löfvendahl, 203–10. Stockholm: ICOMOS Sweden.

Lopez-Arce, P., E. Doehne, J. Greenshields, D. Benavente, and D. Young. 2009. Treatment of rising damp and salt decay: The historic masonry buildings of Adelaide, South Australia. *Materials and Structures/Materiaux et constructions* 42 (6): 827–48.

Lotzmann, S., and H. R. Sasse. 1999. Drilling resistance as an indicator for effectiveness of stone consolidation. In *The Use of and Need for Preservation Standards in Architectural Conservation,* ed. L. B. Sickles-Taves, 77–89. ASTM Special Technical Publication 1355. W. Conshohocken, PA: ASTM.

Luan Xiaoxia, Chunchun Xu, Zise Wang, and Jing Ji. 2008. Preservation effect of modified waterborne epoxy resin emulsion for stone historical relic [in Chinese]. *Fushi yu fanghu = Corrosion & Protection* 29 (8): 451–53.

Lubelli, B., R. P. J. Van Hees, C. Groot, and J. Gunneweg. 2007. Risks of the use of water repellents on salt contaminated masonry: The case of a windmill in the Netherlands. *Restoration of Buildings and Monuments: An International Journal = Bauinstandsetzen und Baudenkmalpflege: Eine internationale Zeitschrift* 13 (5): 319–30.

MacDonald, J., B. Thomson, and K. Tonge. 1992. Chemical cleaning of sandstone: Comparative laboratory studies. In *Stone Cleaning and the Nature, Soiling and Decay Mechanisms of Stone: Proceedings of the International Conference Held in Edinburgh, UK, 14–16 April 1992,* ed. R. G. M. Webster, 217–26. London: Donhead Publishing.

MacLeod, I. 2000. Rock art conservation and management: The past, present and future options. *Reviews in Conservation* 1: 32–45.

MacLeod, I. D., and P. Haydock. 2002. Microclimate modelling for prediction of environmental conditions within rock art shelters. In *13th Triennial Meeting, ICOM Committee for Conservation, Rio de Janiero, 22–27 September 2002: Preprints,* ed. R. Vontobel, 571–77. London: James & James.

MacLeod, I., P. Haydock, D. Tulloch, and B. Ford. 1995. Effects of microbiological activity on the conservation of Aboriginal rock art. *AICCM Bulletin* 21 (1): 3–9. http://www.aiccm.org.au/docs/Bulletin1995/MacLeodEtAl1995Vol21No1.pdf.

Madrid, A., R. Coffman, and W. Ginell. 1993. Air void detection and characterization in thermal insulators using IR thermography. In *8th International Conference on Thermal Engineering and Thermography, Budapest, Hungary, June, 1993: Abstracts,* ed. I. Benko, 225–26. Budapest: Scientific Society of Measurement and Automation, Branch of Thermo Technical Measurements and Thermography.

Maekawa, S. 2006. Comparison of shelters' performance at Joya de Ceren, El Salvador. In *Heritage, Weathering and Conservation: Proceedings of the International Conference on Heritage, Weathering and Conservation (HWC-2006), 21–24 June 2006, Madrid,* ed. R. Fort, M. Alvarez de Buergo, M. Gómez-Heras, and C. Vazquez-Calvo, 469–76. Balkema Proceedings and Monographs in Engineering, Water and Earth Sciences. London and New York: Taylor & Francis.

Maekawa, S., F. Lambert, and J. Meyer. 1995. Environmental monitoring at Tiwanaku. In *Materials Issues in Art and Archaeology IV: Symposium Held May 16–21, 1994, Cancun, Mexico,* ed. P. B. Vandiver, J. R. Druzik, J. L. Galván Madrid, I. C. Freestone, and G. S. Wheeler, 885–92. Materials Research Society Symposium Proceedings 352. Pittsburgh: Materials Research Society.

Mairani, A., M. Matteini, and M. Rizzi. 2000. L'Eterno Padre' di Baccio Bandinelli: Fondamenti e tecniche; Del trattamento di protezione ad 'ossalato di calcio artificiale di una scultura marmorea. *OPD Restauro* 12: 146–50.

Maish, J. P., and E. Risser. 2002. A case study in the use of cyclododecane and latex rubber in the molding of marble. *Journal of the American Institute for Conservation* 41 (2): 127–37.

Malaga, K., B. Schouenborg, and B. Grelk. 2008. Combadura y dilatación de paneles de piedra natural: Ensayo y evaluación de mármol y caliza [Bowing and expansion of natural stone panels: Marble and limestone testing and assessment]. *Materiales de construccion* 58 (289–90): 97–112.

Malaurent, P., J. Brunet, D. Lacanette, and J.-P. Caltagirone. 2007. Contribution of numerical modelling of environmental parameters to the conservation of prehistoric cave paintings: The example of Lascaux Cave. *Conservation and Management of Archaeological Sites* 8 (2): 59–76.

Malzbender, T., D. Gelb, and H. Wolters. 2001. Polynomial texture maps. In *SIGGRAPH '01: Proceedings of the 28th Annual Conference on Computer Graphics and Interactive Techniques,* 519–28. [New York]: Association for Computing Machinery.

Mamillan, M. 1991. Methods of assessing the deterioration of stone monuments. In *Science, Technology, and European Cultural Heritage: Proceedings of the European Symposium, Bologna, Italy, 13–16 June 1989,* ed. N. S. Baer, C. Sabbioni, and A. I. Sors, 90–99. Oxford and Boston: Published for the Commission of the European Communities by Butterworth-Heinemann Publishers.

Mangio, R., M. Simpson, and M. Tokarz. 1996. The influence of sodium sulphate crystallisation on fresh and artificially weathered Salem limestone and Shelburne marble treated with silica sols. In *Conservation et restauration des biens culturels: Pierre, pollution atmosphérique, peinture murale, études scientifiques et cas pratiques; Actes du Congrès LCP 1995, Montreux 24–29 septembre 1995 = Preservation and Restoration of Cultural Heritage: Stone Materials, Air Pollution, Murals, Scientific Research Work and Case Studies: Proceedings of the 1995 LPC Congress, Montreux 24–29 September 1995,* ed. R. Pancella, 249–60. Lausanne: Laboratoire de conservation de la pierre, Département des matériaux, EPFL (Ecole polytechnique fédérale de Lausanne).

Mansch, R., and E. Bock. 1998. Biodeterioration of natural stone with special reference to nitrifying bacteria. *Biodegradation* 9 (1): 47–64.

Mao, L., and D. Kagi. 1995. Method for assessing liquids for the remedial treatment of rising damp. In *Methods of Evaluating Products for the Conservation of Porous Building Materials in Monuments: International Colloquium, Rome, 19–21 June 1995; Preprints,* 359–71. Rome: ICCROM.

Maravelaki, P., G. Biscontin, R. Polloni, W. Cecchetti, and E. Zendri. 1992. Investigation on surface alteration of limestone related to cleaning processes. In *Proceedings of the 7th International Congress on Deterioration and Conservation of Stone, Lisbon, 15–18 June 1992,* ed. J. Delgado Rodrigues, F. Henriques, and F. Telmo Jeremias, 1093–101. Lisbon: Laboratório Nacional de Engenharia Civil.

Maravelaki-Kalaitzaki, P., N. Kallithrakas-Kontos, Z. Agioutantis, S. Maurigiannakis, and D. Korakaki. 2008. A comparative study of porous limestones treated with silicon-based strengthening agents. *Progress in Organic Coatings* 62 (1): 49–60.

Maravelaki-Kalaitzaki, P., V. Zafiropulos, and C. Fotakis. 1999. Excimer laser cleaning of encrustation on Pentelic marble: Procedure and evaluation of the effects. *Applied Surface Science* 148 (1): 92–104.

Margheri, F., S. Modi, L. Masotti, P. Mazzinghi, R. Pini, S. Siano, and R. Salimbeni. 2000. SMART CLEAN: A new laser system with improved emission characteristics and transmission through long optical fibres. *Journal of Cultural Heritage* 1 (Suppl. 1): S119–23.

Mariani, A., P. Capelletti, and A. Brunetti. 2008. Frontal polymerization: A new approach to the consolidation of stone. In *Stone Consolidation in Cultural Heritage: Research and Practice; Proceedings of the International Symposium, Lisbon, 6–7 May 2008,* ed. J. Delgado Rodrigues and J. M. Mimoso, 91–100. Lisbon: LNEC (Laboratório Nacional de Engenharia Civil).

Marini, P., and R. Bellopede. 2009. Bowing of marble slabs: Evolution and correlation with mechanical decay. *Construction and Building Materials* 23 (7): 2599–605.

Martin, B., K. Röller, and B. Stöckhert. 1999. Low-stress pressure solution experiments on halite single-crystals. *Tectonophysics* 308 (3): 299–310.

Martín-Gil, J., F. J. Martín-Gil, M. D. C. Ramos-Sánchez, and P. Martín-Ramos. 2005. The orange-brown patina of Salisbury Cathedral (west porch) surfaces: Evidence of its man-made origin. *Environmental Science and Pollution Research* 12 (5): 285–89.

Maruthamuthu, S., N. Muthukumar, M. Natesan, and N. Palaniswamy. 2008. Role of air microbes on atmospheric corrosion. *Current Science* 94 (3): 359–63.

Maslow, A. H. 2006. *Toward a Psychology of Being.* 3rd ed. Moorpark: Academic Internet Publishers Incorporated.

Massa, S., S. Naldini, and A. Rorro. 1991. A system of classification of the decay of stone monuments. In *Science, Technology, and European Cultural Heritage: Proceedings of the European Symposium, Bologna, Italy, 13–16 June 1989,* ed. N. S. Baer, C. Sabbioni, and A. I. Sors, 859–62. Oxford and Boston: Published for the Commission of the European Communities by Butterworth-Heinemann Publishers.

Massey, S. W. 1999. The effects of ozone and NO(x) on the deterioration of calcareous stone. *Science of the Total Environment* 227 (2–3): 109–21.

Matsuoka, N., and J. Murton. 2008. Frost weathering: Recent advances and future directions. *Permafrost and Periglacial Processes* 19 (2): 195–210.

Matteini, M. 1991. In review: An assessment of Florentine methods of wallpainting conservation based on the use of mineral treatments. In *The Conservation of Wall Paintings: Proceedings of a Symposium Organized by the Courtauld Institute of Art and the Getty Conservation Institute, London, July 13–16, 1987,* ed. S. Cather, 137–48. Marina del Rey, CA: Getty Conservation Institute. http://www.getty.edu/conservation/publications/pdf_publications/wall_paintings.pdf.

Matteini, M., and P. Zannini. 2004. Mineral treatments for the conservation of cultural heritage materials: Analytical characterization and trial applications of barium oxalates and aluminates. *Science and Technology for Cultural Heritage* 13 (1–2): 11–18.

Matteini, M., A. Moles, and S. Giovannoni. 1994. Calcium oxalate as a protective mineral system for wall paintings: Methodology and analyses. In *Conservation of Monuments in the Mediterranean Basin: Stone Monuments, Methodologies for the Analysis of Weathering and Conservation; Proceedings of the 3rd International Symposium, Venice, 22–25 June 1994 = La conservazione dei monumenti nel bacino del Mediterraneo: Materiali lapidei e monumenti, metodologie per l'analisi del degrado e la conservazione; Atti del 3° simposio internazionale, Venezia, 22–25 giugno 1994,* ed. V. Fassina, H. Ott, and F. Zezza, 155–62. Venice: Soprintendenza ai beni artistici e storici di Venezia.

Maxwell, I. 1992. Stone cleaning: For better or worse? An overview. In *Stone Cleaning and the Nature, Soiling and Decay Mechanisms of Stone: Proceedings of the International Conference Held in Edinburgh, UK, 14–16 April 1992,* ed. R. G. M. Webster, 3–49. London: Donhead Publishing.

Maxwell, I. 1996. Emerging conservation issues in consequence of cleaning Scottish historic buildings. In *Proceedings of the 8th International Congress on Deterioration and Conservation of Stone, Berlin, 30 Sept.–4 Oct. 1996,* ed. J. Riederer, 1405–14. Berlin: Möller Druck und Verlag.

Maxwell, I. 2007. *Cleaning Sandstone: Risks and Consequences*. Inform: Information for Historic Building Owners. Edinburgh: Historic Scotland, Technical Conservation, Research and Education Group.

May, E., F. J. Lewis, S. Pereira, S. Tayler, M. R. D. Seaward, and D. Allsopp. 1993. Microbial deterioration of building stone: A review. *Biodeterioration Abstracts* 7 (2): 109–23.

May, E., A. Webster, C. Krage, L. Sorlini, A. Vgenopoulos, E. Mello, and J. Kuever. 2004. Databases for stone conservation by bioremediation arising from a European collaboration: The Biobrush Project. In *Proceedings of the 10th International Congress on Deterioration and Conservation of Stone, Stockholm, June 27–July 2, 2004*, ed. D. Kwiatkowski and R. Löfvendahl, vol. 2, 753–60. Stockholm: ICOMOS Sweden.

May, E., A. M. Webster, R. Inkpen, D. Zamarrefio, J. Kuever, C. Rudolph, T. Warscheid, C. Sorlini, F. Cappitelli, E. Zanardini, G. Ranalli, L. Krage, A. Vgenopoulos, D. Katsinis, E. Mello, and M. Malagodi . 2008. The Biobrush Project for bioremediation of heritage stone. In *Heritage Microbiology and Science: Microbes, Monuments and Maritime Materials*, ed. E. May, M. Jones, and J. Mitchell, 76–93. Royal Society of Chemistry (Great Britain) Special Publication 315. Cambridge: Royal Society of Chemistry. http://www.rsc.org/ebooks/2008/BK9780854041411/BK9780854041411-00076.pdf.

Mazzeo, R., and B. Eshøj. 2002. Designing university postgraduate curricula for conservation scientists. In *13th Triennial Meeting, ICOM Committee for Conservation, Rio de Janiero, 22–27 September 2002: Preprints*, ed. R. Vontobel, 137–41. London: James & James.

Mazzeo, R., and B. Eshøj. 2008. Education of conservation scientists: The EPISCON project; European PhD in science for conservation. In *15th Triennial Conference, ICOM Committee for Conservation, New Delhi, 22–26 September 2008: Preprints*, ed. J. Bridgland, 116–21. New Delhi: Allied Publishers.

Mazzinghi, P., and F. Margheri. 2003. A short pulse, free running, Nd: YAG laser for the cleaning of stone cultural heritage. *Optics and Lasers in Engineering* 39 (2): 191–202.

McAlister, J. J., B. J. Smith, and A. Török. 2008. Transition metals and water-soluble ions in deposits on a building and their potential catalysis of stone decay. *Atmospheric Environment* 42 (33): 7657–68.

McCabe, S., J. M. McKinley, and B. J. Smith. 2008. Simulating initial stages of salt accumulation and organisation within building sandstones. In *Proceedings of the 11th International Congress on Deterioration and Conservation of Stone, 15–20 September 2008, Toruń, Poland*, ed. J. W. Lukaszewicz and P. Niemcewicz, 173–80. Toruń, Poland: Nicolaus Copernicus University.

McLane, J., R. A. J. Wüst, B. Porter, and J. Rutherford. 2003. Flash-flood impacts and protection measures in the Valley of the Kings, Luxor, Egypt. *APT Bulletin* 34 (1): 37–45.

McNamara, C. J., T. D. Perry IV, K. Bearce, G. Hernandez-Duque, and R. Mitchell. 2005. Measurement of limestone biodeterioration using the Ca^{2+} binding fluorochrome Rhod-5N. *Journal of Microbiological Methods* 61 (2): 245–50.

McNamara, C. J., T. D. Perry IV, K. A. Bearce, G. Hernandez-Duque, and R. Mitchell. 2006. Epilithic and endolithic bacterial communities in limestone from a Maya archaeological site. *Microbial Ecology* 51 (1): 51–64.

Meierding, T. C. 1993. Marble tombstone weathering and air pollution in North America. *Annals of the Association of American Geographers* 83 (4): 568–88.

Meierding, T. 2000. Philadelphia's effect on precipitation acidity from marble gravestone dissolution rates. *Pennsylvania Geographer* 38 (1): 42–56.

Meiklejohn, K. I. 1995. The deterioration and preservation of rock art in the KwaZulu/Natal Clarens Formation: A geomorphological perspective. *Pictogram* 8: 1–13.

Meiklejohn, K. I. 1997. The role of moisture in the weathering of the Clarens Formation of the KwaZulu-Natal Drakensberg: Implications for the preservation of indigenous rock art. *South African Geographical Journal* 79 (3): 199–206.

Meiklejohn, K. I., K. Hall, and J. K. Davis. 2009. Weathering of rock art at two sites in the KwaZulu-Natal Drakensberg, southern Africa. *Journal of Archaeological Science* 36 (4): 973–79.

Meinhardt-Degen, J., and R. Snethlage. 2007. Investigations on the effect of a re-treatment of sandstone using Greensandstone of Regensburg and Green Mainsandstone as an example = Materialtechnische Untersuchungen zur Risikoabschätzung von Folgekonservierungen bei Sandsteinen am Beispiel von Regensburger Grünsandstein und Grünem Mainsandstein. *Zeitschrift der Deutschen Gesellschaft für Geowissenschaften* 158 (4): 921–29.

Meinlschmidt, P., T. Bothe, K. D. Hinsch, and L. Mehlhorn. 1998. Nondestructive testing and evaluation of historical monuments using thermography and electronic speckle pattern interferometry (ESPI). In *Nondestructive Evaluation of Materials and Composites II, 31 March–1 April 1998, San Antonio, Texas,* ed. S. R. Doctor, C. A. Lebowitz, and G. Y. Baaklini, 28–36. Proceedings of SPIE (The International Society for Optical Engineering) 3396. Bellingham, WA: SPIE.

Meinlschmidt, P., G. Gulker, K. Hinsch, C. Holscher, J. Delgado Rodrigues, F. Henriques, and F. T. Jeremias. 1992. Nondestructive microdeformation-measurements on stone surfaces with video holography. In *Proceedings of the 7th International Congress on Deterioration and Conservation of Stone, Lisbon, 15–18 June 1992,* ed. J. Delgado Rodrigues, F. Henriques, and F. Telmo Jeremias, 937–46. Lisbon: Laboratório Nacional de Engenharia Civil.

Mérindol, P. 1994. Techniques de nettoyage et de consolidation utilisées pour la restauration du portail de l'église saint-trophime d'arles. In *EUROCARE-EUROMARBLE EU 496: Proceedings of the Fourth Workshop, Arles, France, November 3–6, 1993,* ed. R. Snethlage and V. Vergès-Belmin, 93–98. Forschungsbericht (Bayerisches Landesamt für Denkmalpflege) 13. Munich: Bayerisches Landesamt für Denkmalpflege.

Mertz, J.-D., J.-L. Grunenwald, and J. D. Ternay. 2003. Traitements préventifs anti-graffiti: Application aux matériaux du patrimoine bâti nantais. *Pierre actual: Matériaux, ouvrages, techniques* 801: 68–79.

Metaxa, E., T. Agelakopoulou, I. Bassiotis, Ch. Karagianni, and F. Roubani-Kalantzopoulou. 2009. Gas chromatographic study of degradation phenomena concerning building and cultural heritage materials. *Journal of Hazardous Materials* 164 (2–3): 592–99.

Miget, S. 2000. Nettoyage à sec par polymerisation d'un film plastique. *Les cahiers techniques du bâtiment* (211): 46–47, 49.

Miliani, C., M. L. Velo-Simpson, and G. W. Scherer. 2007. Particle-modified consolidants: A study on the effect of particles on sol-gel properties and consolidation effectiveness. *Journal of Cultural Heritage* 8 (1): 1–6.

Ministère de la culture et de la communication. 2003. Ouvrages en pierre de taille : fascicule technique, cahier des clauses techniques particulières, mode de métré et bordereau de prix unitaire. Paris : Sous direction des monuments historiques, mission études et travaux, 41 and 63.

Miquel, A., P. Bromblet, V. Vergès-Belmin, L. Binda, G. Baronio, E. De Witte, H. De Clercq, R. Van Hees, and H. Brocken. 2001. Experiments on the compatibility of a polysiloxane treatment with substrates loaded with sodium sulphate: Influence of the physical properties of substrates on the salt content limit. In *Surface Technology With Water Repellent Agents: Proceedings of Hydrophobe III, Third International Conference on Surface Technology With Water Repellent Agents, Universität Hannover, Germany, September 25 and 26, 2001,* ed. K. Littmann and A. E. Charola, 203–17. Freiburg: Aedificatio Verlag.

Miquel, A., P. Bromblet, V. Vergès-Belmin, L. Binda, G. Baronio, E. De Witte, H. De Clercq, R. Van Hees, and H. Brocken. 2002. Experimental study on the compatibility of a polysiloxane treatment with substrates loaded with sodium sulphate: Influence of the physical properties of the substrates on the salt content limit. *Restoration of Buildings and Monuments: An International Journal = Bauinstandsetzen und Baudenkmalpflege: Eine internationale Zeitschrift* 8 (2–3): 271–91.

Mirowski, R. 1988. A new method of impregnation of stone historical objects. In *Proceedings of the 6th International Congress on Deterioration and Conservation of Stone, Toruń, 12–14.09.1988,* ed. J. Ciabach, 633–40. Toruń, Poland: Nicolaus Copernicus University.

Mitchell, D. J., and D. E. Searle, eds. 2004. *Stone Deterioration in Polluted Urban Environments.* Land Reconstruction and Management Series 3. Enfield, NH: Science Publishers.

Mohammadi, P., and W. E. Krumbein. 2008. Biodeterioration of ancient stone materials from the Persepolis monuments (Iran). *Aerobiologia* 24 (1): 27–33.

Molaro, J. L., and C. P. McKay. 2010. Processes controlling rapid temperature variations on rock surfaces. *Earth Surface Processes and Landforms* 35. DOI: 10.1002/esp.1957.

Monte, M. 2003. Oxalate film formation on marble specimens caused by fungus. *Journal of Cultural Heritage* 4 (3): 255–58.

Monte, M., R. Ferrari, G. Lonati, and M. Malagodi. 2000. Biocidal activity on microbic bio-deteriogens and on frescoes. In *2nd International Congress on Science and Technology for the Safeguard of Cultural Heritage in the Mediterranean Basin: 5–9 July 1999, Paris: Proceedings,* ed. A. Guarino, vol. 1, 633–39. Paris: Éd. scientifiques et médicales Elsevier.

Montoto, M., L. Valdeón, and R. M. Esbert. 1996. Non-destructive tests in stone conservation: Tomography of the Axeitos (La Coruña, Spain) Megalith. In *Degradation and Conservation of Granitic Rocks in Monuments: Proceedings of the EC Workshop Held in Santiago de Compostela (Spain) on 28–30 November 1994,* ed. M. A. Vicente Hernández, J. Delgado Rodrigues, and J. Acevedo, 281–87. Protection and Conservation of the European Cultural Heritage Research Report 5. Brussels: European Commission, Directorate-General XII, Science, Research and Development.

Moore, S., M. J. Beazley, M. R. McCallum, and J. Russ. 2000. Can calcium oxalate residues from lichen activity reflect past climate change? In *Preprints of Extended Abstracts Presented at the 220th ACS National Meeting, Washington, DC, August 20–24, 2000,* ed. M. L. Trehy, 4–5. Preprints of Extended Abstracts, vol. 40, no. 2. [United States]: American Chemical Society, Division of Environmental Chemistry. http://www.envirofacs.org/Pre-prints/Vol%2040%20No%202/General/Sec%20A/p209.PDF.

Moraes Rodrigues, D. de, and S. D. Emery. 2008. Revisiting the conservation of the flooring of the Great Stone Church: Results of the use of ethyl silicate five years after treatment. In *Stone Consolidation in Cultural Heritage: Research and Practice; Proceedings of the International Symposium, Lisbon, 6–7 May 2008,* ed. J. Delgado Rodrigues and J. M. Mimoso, 241–50. Lisbon: LNEC (Laboratório Nacional de Engenharia Civil).

Morasset, E. V. 2008. Cleaning historic building interiors: the question of residue using Arte Mundit® cleaning paste. Master's thesis, Columbia University.

Morasset, E. V., A. Rizzo, J. Arslanoglu, and G. Wheeler. 2009. Cleaning historic building interiors: The question of residue using Arte Mundit® cleaning paste. In *2009 Architecture Specialty Group Abstracts: AIC Annual Meeting, Conservation 2.0—New Directions, Los Angeles, May 19–22, 2009*. Washington, DC: American Institute for Conservation of Historic and Artistic Works. http://www.conservation-us.org/index.cfm?fuseaction=page.viewpage&pageid=840.

Moreau, C. 2008. Vieillissement naturel en milieu urbain de pierres calcaires hydrofugées: Évaluation de la durabilité des traitements et de leur impact sur le nettoyage. PhD diss., Université de Reims Champagne-Ardenne.

Moreau, C., V. Vergès-Belmin, L. Leroux, G. Orial, G. Fronteau, and V. Barbin. 2008. Water-repellent and biocide treatments: Assessment of the potential combinations. *Journal of Cultural Heritage* 9 (4): 394–400.

Moropoulou, A., N. P. Avdelidis, and P. Theoulakis. 2003. Outdoor thermographic survey of historic structures. In *Thermosense XXV, 1 April 2003,* ed. K. E. Cramer and X. P. Maldague, 117–21. Proceedings of SPIE (The International Society for Optical Engineering) 5073. Bellingham, WA: SPIE.

Moropoulou, A., P. Theoulakis, T. Tsiourva, and G. Haralampopoulos. 2000. Compatibility evaluation of consolidation treatments in monument scale: Techniques and methodology to evaluate the compatibility of conservation materials and interventions in monument scale = Evaluation de compatibilité des traitements de consolidation à l'échelle de monuments. In *Compatible Materials Recommendations for the Preservation of European Cultural Heritage,* ed. A. Moropoulou, 209–30. PACT 59. Athens: Association of Civil Engineers of Greece.

Morris, W. 1877. Manifesto of the Society for the Protection of Ancient Buildings. Available at http://www.spab.org.uk/what-is-spab-/the-manifesto/.

Mosch, S., and S. Siegesmund. 2007. Petrophysical and technical properties of dimensional stones: A statistical approach = Statistisches Verhalten petrophysikalischer und technischer Eigenschaften von Naturwerksteinen. *Zeitschrift der Deutschen Gesellschaft für Geowissenschaften* 158 (4): 821–68.

Mosquera, M. J., and D. M. de los Santos. 2008. New water-repellent nanomaterial for protecting and consolidating stone. In *Proceedings of the 11th International Congress on Deterioration and Conservation of Stone, 15–20 September 2008, Toruń, Poland,* ed. J. W. Lukaszewicz and P. Niemcewicz, 993–1000. Toruń, Poland: Nicolaus Copernicus University.

Mossotti, V. G., and L. M. Castanier. 1990. The measurement of water transport in Salem limestone by X-ray computer aided tomography. In *The Engineering Geology of Ancient Works, Monuments and Historical Sites: Preservation and Protection; Proceedings of an International Symposium Organized by the Greek National Group of IAEG, Athens, 19–23 September 1988,* ed. P. G. Marinos and G. C. Koukis, vol. 4, 2079–82. Rotterdam and Brookfield, VT: Balkema.

Muros, V., and J. Hirx. 2004. The use of cyclododecane as a temporary barrier for water-sensitive ink on archaeological ceramics during desalination. *Journal of the American Institute for Conservation* 43 (1): 75–89.

Mustoe, G. E. 2010. Biogenic origin of coastal honeycomb weathering. *Earth Surface Processes and Landforms* 35: 424–34.

Nasraoui, M., W. Nowik, and B. Lubelli. 2009. A comparative study of hygroscopic moisture content, electrical conductivity and ion chromatography for salt assessment in plasters of historical buildings. *Construction and Building Materials* 23 (5): 1731–35.

Niesewand, N. 1999. British Museum faces inquiry into use of "wrong kind of stone." *The Independent (London, England)*, October 22.

Nijland, T. G. and T. J. Wijffels. 2003. Laser cleaning of Rakowicze sandstone. *Heron* 48(3): 197–205.

Nimmrichter, J., and R. Linke. 2008. Evaluation of hydrophobisation on compact limestone and calacareous tuff with negative result: Case studies of the facades of the Cathedral of Salzburg and Gothic churches in Upper Austria. In *Proceedings of the 11th International Congress on Deterioration and Conservation of Stone, 15–20 September 2008, Toruń, Poland,* ed. J. W. Lukaszewicz and P. Niemcewicz, 1019–25. Toruń, Poland: Nicolaus Copernicus University.

Nocita, B. W. 1987. Thermal expansion of salts as an effective physical weathering process in hot desert environments. In *Abstracts with Programs: Geological Society of America,* vol. 19, 790. Boulder, CO: Geological Society of America.

Nord, A. G., and K. Holenyi. 1999. Sulphur deposition and damage on limestone and sandstone in Stockholm city buildings. *Water, Air, and Soil Pollution* 109 (1–4): 147–62.

Normandin, K. C., L. Powers, D. Slaton, and M. J. Scheffler. 2007. The potential use of laser ablation for selective cleaning of Indiana limestone. In *Lasers in the Conservation of Artworks: LACONA VI, Vienna, Austria, Sept. 21–25, 2005; Proceedings,* ed. J. Nimmrichter, W. Kautek, and M. Schreiner, 65–73. Springer Proceedings in Physics 116. Berlin and New York: Springer.

Normandin, K. C., D. Slaton, N. R. Weiss, and J. Pearce, eds. 2005. Cleaning Techniques in Conservation Practice. Special issue, *Journal of Architectural Conservation* 11 (3).

Norwick, S. A., and L. R. Dexter. 2002. Rates of development of tafoni in the Moenkopi and Kaibab formations in Meteor Crater and on the Colorado Plateau, northeastern Arizona. *Earth Surface Processes and Landforms* 27 (1): 11–26.

Odgers, D. 2003. Cleaning St Paul's: Conservation technology and practice. In *Historic Churches: The Conservation and Repair of Ecclesiastical Buildings,* 10th ed., 14–16. Building Conservation Directory Special Report Magazine. Tisbury: Cathedral Communications.

Odgers, D. 2006. A conservator's role: Personal passion and public perception. *Icon News* (2): 24–26.

Orial, G., and F. Bousta. 2005. Les altérations biologiques et les biens patrimoniaux: Les traitements: Définitions, sélection des produits et mise en oeuvre.

Monumental: Revue scientifique et technique de la Sous-direction des monuments historiques 2005 (1): 107–12.

Orial, G., and A. Brunet. 1992. Elne: The old cathedral cloister; The capitals up against a bacterial attack. In *La conservation des monuments dans le bassin méditerranéen: Actes du 2ème symposium international = The Conservation of Monuments in the Mediterranean Basin: Proceedings of the 2nd International Symposium,* ed. D. Decrouez, J. Chamay, and F. Zezza, 542–49. Geneva: Ville de Genève, Muséum d'histoire naturelle, and Musée d'art et d'histoire.

Orial, G., and A. Brunet. 2004. The Rennes Parliament in France: Controlling a generalised contamination. In *Schimmel: Gefahr für Mensch und Kulturgut durch Mikroorganismen = Fungi: A Threat for People and Cultural Heritage through Micro-Organisms,* ed. A. Rauch, S. Miklin-Kniefacz, and A. Harmssen, 121–30. VDR Schriftenreihe 1. Bonn: VDR (Verband der Restauratoren); Stuttgart: K. Theiss.

Orial, G., and G. Riboulet. 1993. Technique de nettoyage de la statuaire monumentale par désincrustation photonique: Réalisation d'un prototype mobile. In *Conservation of Stone and Other Materials: Proceedings of the International RILEM/UNESCO Congress "Conservation of Stone and Other Materials: Research – Industry – Media," UNESCO Headquarters, Paris, June 29–July 1, 1993,* ed. M.-J. Thiel, 542–49. RILEM Proceedings 21. London and New York: E & FN Spon.

Orial, G., T. Vieweger, and J. F. Loubiere. 2003. Biological mortars: A solution for stone sculpture conservation. In *Art, Biology, and Conservation: Biodeterioration of Works of Art,* ed. R. J. Koestler, V. H. Koestler, A. E. Charola, and F. E. Nieto-Fernandez, 498–517. New York: Metropolitan Museum of Art.

Orial, G., S. Castanier, G. Levrel, and J.-F. Loubière. 1996. Les bactéries architectes = Builder bacteria. *Coré: Conservation et restauration du patrimoine culturel* 1: 58–62.

Ortega-Calvo, J. J., X. Arino, M. Hernandez-Marine, and C. Saiz-Jimenez. 1995. Factors affecting the weathering and colonization of monuments by phototrophic microorganisms. *Science of the Total Environment* 167: 329–41.

Österreichisches Normungsinstitut (ON). 2006. *ÖNORM B 3355-1: Trockenlegung von feuchtem Mauerwerk—Teil 1: Bauwerksdiagnose und Planungsgrundlagen.* Berlin (Deutschland, Bundesrepublik): Beuth Verlag GmbH.

Oudbashi, O., B. Siapoosh, M. Siapoosh, and A. Shekofte. 2008. Consolidation of limestone by the lime method in Sangneveshteh Historic Monument, Khoramabad, west of Iran: A practical case. In *Stone Consolidation in Cultural Heritage: Research and Practice; Proceedings of the International Symposium, Lisbon, 6–7 May 2008,* ed. J. Delgado Rodrigues and J. M. Mimoso, 289–98. Lisbon: LNEC (Laboratório Nacional de Engenharia Civil).

Padfield, J., D. Saunders, and T. Malzbender. 2005. Polynomial texture mapping: A new tool for examining the surface of paintings. In *14th Triennial Meeting, ICOM Committee for Conservation, The Hague, 12–16 September 2005: Preprints,* ed. I. Verger, 504–10. London: James & James, Earthscan.

Padgett, A., and K. Barthuli. 1997. Rock art documentation and assessment. *CRM* 20 (10): 28–30.

Palieraki, V., E. Vintzileou, A. Miltiadou-Fezans, and N. Delinikolas. 2008. The use of radar techniques and boroscopy in investigating old masonry: The case of Dafni Monastery. *International Journal of Architectural Heritage* 2 (2): 155–86.

Palmer, A. 2002. The cleaning power of Pisa: The colour has drained from Pisa's medieval frescoes and the restorers stand accused. *The Daily Telegraph*, May 14. http://www.telegraph.co.uk/culture/art/3577408/The-cleaning-power-of-Pisa.html.

Pamplona, M., M. Kocher, R. Snethlage, and L. Aires-Barros. 2008. Drilling resistance: State of the art and future developments. In *Proceedings of the 11th International Congress on Deterioration and Conservation of Stone, 15–20 September 2008, Toruń, Poland,* ed. J. W. Lukaszewicz and P. Niemcewicz, 449–56. Toruń, Poland: Nicolaus Copernicus University.

Papida, S., W. Murphy, and E. May. 2000. Enhancement of physical weathering of building stones by microbial populations. *International Biodeterioration and Biodegradation* 46 (4): 305–17.

Paradise, T. R. 2002. Sandsteinverwitterung und ausprägung in Petra, Jordanien [Sandstone weathering and aspect in Petra, Jordan]. *Zeitschrift für Geomorphologie* 46 (1): 1–17.

Paradise, T. R. 2005. Petra revisited: An examination of sandstone weathering research in Petra, Jordan. In *Stone Decay in the Architectural Environment,* ed. A. V. Turkington, 39–49. GSA Special Paper 390. Boulder, CO: Geological Society of America.

Paterakis, A. B. 1999. Those evasive salt crystals. In *12th Triennial Meeting, ICOM Committee for Conservation, Lyon, 29 August–3 September 1999: Preprints,* ed. J. Bridgland, 799–802. London: James & James (Science Publishers) Ltd.

Pearson, C., and J. Clarke. 1978. Conservation and restoration of painting and engraving sites in Western Australia. In *Conservation of Rock Art: Proceedings of the International Workshop on the Conservation of Rock Art, Perth, September 1977,* ed. C. Pearson, 89–94. Sydney: Institute for the Conservation of Cultural Material.

Pel, L., H. Huinink, and K. Kopinga. 2002. Ion transport and crystallization in inorganic building materials as studied by nuclear magnetic resonance. *Applied Physics Letters* 81 (15): 2893.

Pel, L., K. Kopinga, and H. Brocken. 1996. Moisture transport in porous building materials. *Heron* 41 (2): 95–105.

Pel, L., A. Sawdy, and V. Voroninaa. 2010. Physical principles and efficiency of salt extraction by poulticing. *Journal of Cultural Heritage* 11 (1): 59–67.

Pepi, R. M. 2008. Book review: Stone Conservation Principles and Practice. *Traditional Building* (December). http://www.traditional-building.com/Previous-Issues-08/DecemberBR08Stone.html.

Pérez Bernal, J. L., and M. A. Bello López. 2000. Fractal dimension of stone pore surface as weathering descriptor. *Applied Surface Science* 161 (1): 47–53.

Perry, T. D., IV, O. W. Duckworth, C. J. McNamara, S. T. Martin, and R. Mitchell. 2004. Effects of the biologically produced polymer alginic acid on macroscopic and microscopic calcite dissolution rates. *Environmental Science and Technology* 38 (11): 3040–46.

Perry, T. D., IV, C. J. McNamara, and R. Mitchell. 2005. Biodeterioration of stone. In *Scientific Examination of Art: Modern Techniques in Conservation and Analysis; Sackler National Academy of Sciences Colloquium,* 72–86. Washington, DC: National Academies Press. http://books.nap.edu/openbook.php?record_id=11413&page=72.

Petre, J. 2006. Crumbling cathedral held together by tape. *The Daily Telegraph* (UK), October 4. http://www.telegraph.co.uk/news/uknews/1530481/Crumbling-cathedral-held-together-by-tape.html.

Petushkova, Y. P., and A. F. Grishkova. 1990. Bacterial degradation of limestone treated with polymers. In *9th Triennial Meeting, ICOM Committee for Conservation, Dresden, German Democratic Republic, 26–31 August, 1990: Preprints,* ed. K. Grimstad, 347–49. Los Angeles: ICOM Committee for Conservation; Marina del Rey, CA: Getty Conservation Institute.

Piacenti, F., M. Camaiti, T. Brocchi, and A. Scala. 1993a. New developments in perfluorinated protective agents for stone. In *Conservation of Stone and Other Materials: Proceedings of the International RILEM/UNESCO Congress "Conservation of Stone and Other Materials: Research—Industry—Media," held at UNESCO Headquarters, Paris, June 29–July 1, 1993,* ed. M.-J. Thiel, 733–39. RILEM Proceedings 21. London and New York: E & FN Spon.

Piacenti, F., M. Camaiti, C. Manganelli del Fà, and A. Scala. 1993. Fluorinated aggregating materials for stone. In *Conservation of Stone and Other Materials: Proceedings of the International RILEM/UNESCO Congress "Conservation of Stone and Other Materials: Research—Industry—Media," held at UNESCO Headquarters, Paris, June 29–July 1, 1993,* ed. M.-J. Thiel, 740–47. RILEM Proceedings 21. London and New York: E & FN Spon.

Pien, A. 1991. Surface water repellents: Unification of tests methods. In *Science, Technology, and European Cultural Heritage: Proceedings of the European Symposium, Bologna, Italy, 13–16 June 1989,* ed. N. S. Baer, C. Sabbioni, and A. I. Sors, 646–48. Oxford and Boston: Published for the Commission of the European Communities by Butterworth-Heinemann Publishers.

Piervittori, R., O. Salvadori, and D. Isocrono. 2004. Literature on lichens and biodeterioration of stonework, IV. *Lichenologist* 36 (2): 145–57.

Pini, R., S. Siano, and R. Salimbeni. 2000. In field tests and operative applications of improved laser techniques for stone cleaning. In *Proceedings of the 9th International Congress on Deterioration and Conservation of Stone, Venice, June 19–24, 2000,* ed. V. Fassina, vol. 2, 577–82. Amsterdam and New York: Elsevier.

Pinto, A. P. F., and J. Delgado Rodrigues. 2008a. Hydroxlyating conversion treatment and alkoxysilane coupling agent as pre-treatment for the consolidation of limestones with ethly silicate. In *Stone Consolidation in Cultural Heritage: Research and Practice; Proceedings of the International Symposium, Lisbon, 6–7 May 2008,* ed. J. Delgado Rodrigues and J. M. Mimoso, 131–40. Lisbon: LNEC (Laboratório Nacional de Engenharia Civil).

Pinto, A. P. F., and J. Delgado Rodrigues. 2008b. Stone consolidation: The role of treatment procedures. *Journal of Cultural Heritage* 9 (1): 38–53.

Pinto Guerra, E. 2008. *Risanamento di murature umide e degradate: Manuale, guida progettuale, soluzioni.* Palermo: D. Flaccovio.

Poupeleer, A. S., J. Carmeliet, S. Roels, and D. Van Gemert. 2006. Combining expansion/shrinkage monitoring and X-ray measurement for water and salt transport in calcium silicate beam. In *Research in Building Physics and Building Engineering: Proceedings of the 3rd International Building Physics Conference, Concordia University, Montreal, Canada, 27–31 August 2006,* ed. P. Fazio, Hua Ge, Jiwu Rao, and G. Desmarais, 147–54. London and New York: Taylor & Francis.

Price, C. A. 1982. The evaluation of stone preservatives. In *Conservation of Historic Stone Buildings and Monuments: Report of the Committee on Conservation of Historic Stone Buildings and Monuments, National Materials Advisory Board, Commission on Engineering and Technical Systems, National Research Council,* 329–40. Washington, DC: National Academy Press.

Price, C., ed. 2000. *An Expert Chemical Model for Determining the Environmental Conditions Needed to Prevent Salt Damage in Porous Materials: Protection and Conservation of the European Cultural Heritage.* Research Report (European Commission, Directorate-General XII, Science, Research, and Development) 11. London: Archetype Publications.

Price, C. 2006. Consolidation. In *Stone Conservation: Principles and Practice,* ed. A. Henry, 101–26. Shaftesbury, UK: Donhead Publishing.

Price, C. 2007. Predicting environmental conditions to minimise salt damage at the Tower of London: A comparison of two approaches. *Environmental Geology* 52 (2): 369–74.

Price, C., and P. Brimblecombe. 1994. Preventing salt damage in porous materials. In *Preventive Conservation Practice, Theory and Research: Preprints of the Contributions to the Ottawa Congress, 12–16 September 1994,* ed. A. Roy and P. Smith, 90–93. London: International Institute for Conservation of Historic and Artistic Works.

Price, C., K. Ross, and G. White. 1988. A further appraisal of the "lime technique" for limestone consolidation, using a radioactive tracer. *Studies in Conservation* 33 (4): 178–86.

Přikryl, R. 2007. Understanding the earth scientist's role in the pre-restoration research of monuments: An overview. In *Building Stone Decay: From Diagnosis to Conservation,* ed. R. Přikryl and B. J. Smith, 9–21. Geological Society Special Publication 271. London: Geological Society of London.

Přikryl, R., and B. J. Smith. 2007. Diagnosing decay: The value of medical analogy in understanding the weathering of building stones. In *Building Stone Decay: From Diagnosis to Conservation,* ed. R. Přikryl and B. J. Smith, vol. 1, 1–8. Geological Society Special Publication 271. London: Geological Society of London.

Přikryl, R., J. Svobodová, K. Žák, and D. Hradil. 2004. Anthropogenic origin of salt crusts on sandstone sculptures of Prague's Charles Bridge (Czech Republic): Evidence of mineralogy and stable isotope geochemistry. *European Journal of Mineralogy* 16 (4): 609–18.

Pringle, H. 2008. The recent troubles at Lascaux. *Archaeology Magazine Weekly Blog* (August 1). http://archaeology.org/blog/?p=29.

Proietti, N., D. Capitani, S. Cozzolino, M. Valentini, E. Pedemonte, E. Princi, S. Vicini, and A. L. Segre. 2006. In situ and frontal polymerization for the consolidation of porous stones: A unilateral NMR and magnetic resonance imaging study. *Journal of Physical Chemistry B* 110 (47): 23719–28.

Prokos, P., and F. Bala'awi. 2008. Salt weathering in the coastal environment: A thermodynamic approach. In *Proceedings of the 11th International Congress on Deterioration and Conservation of Stone, 15–20 September 2008, Toruń, Poland,* ed. J. W. Lukaszewicz and P. Niemcewicz, 233–41. Toruń, Poland: Nicolaus Copernicus University.

Prudon, Th. H. M. 1975. Book review: The Pattern of English Building by Alec Clifton-Taylor. *Bulletin of the Association for Preservation Technology* 7 (4): 123–24.

Puehringer, J., and L. Engström. 1985. Unconventional methods for the prevention of salt damage. In *Ve Congrès international sur l'altération et la conservation de la pierre: Actes, Lausanne, 25–27.9.1985 = 5th International Congress on Deterioration and Conservation of Stone: Proceedings, Lausanne, 25–27.9.1985,* ed. G. Félix, 241–50. International Congress on Deterioration and Conservation of Stone 5. Lausanne: Presses polytechniques romandes.

Pummer, Erich. 2008. Vacuum-circulation-process: Innovative stone conservation. In *Proceedings of the 11th International Congress on Deterioration and Conservation of Stone, 15–20 September 2008, Toruń, Poland,* ed. J. W. Lukaszewicz and P. Niemcewicz, 481–88. Toruń, Poland: Nicolaus Copernicus University.

Qiang Liu, Bingjian Zhang, Zhongyue Shen, and Huanming Lu. 2006. A crude protective film on historic stones and its artificial preparation through biomimetic synthesis. *Applied Surface Science* 253 (5 [December]): 2625–32.

Qinghua Sun, Aixia Wang, Ximei Jin, Alex Natanzon, Damon Duquaine, Robert D. Brook, Juan-Gilberto S. Aguinaldo, Zahi A. Fayad, Valentin Fuster, Morton Lippmann, Lung Chi Chen, and Sanjay Rajagopalan. 2005. Long-term air pollution exposure and acceleration of atherosclerosis and vascular inflammation in an animal model. *Journal of the American Medical Association* 294 (23): 3003–10.

Ranalli, G., M. Chiavarini, V. Guidetti, F. Marsala, M. Matteini, E. Zanardini, and C. Sorlini. 1996. The use of microorganisms for the removal of nitrates and organic substances on artistic stoneworks. In *Proceedings of the 8th International Congress on Deterioration and Conservation of Stone, Berlin, 30 Sept.–4 Oct. 1996,* ed. J. Riederer, 1415–20. Berlin: Möller Druck und Verlag.

Ranalli, G., M. Matteini, I. Tosini, E. Zanardini, and C. Sorlini. 2000. Bioremediation of cultural heritage: Removal of sulphates, nitrates and organic substances. In *Of Microbes and Art: The Role of Microbial Communities in the Degradation and Protection of Cultural Heritage,* ed. O. Ciferri, P. Tiano, and G. Mastromei, 231–45. New York: Kluwer Academic/Plenum Publishers.

Reddy, M. 1990. Acid rain and air pollution effects on carbonate-stone: Dissolution-runoff experiments. In *Conservation of Monuments in the Mediterranean Basin: The Influence of Coastal Environment and Salt Spray on Limestone Marble; Proceedings of the 1st International Symposium, Bari, 7–10 June 1989 = La conservazione dei monumenti nel bacino del Mediterraneo: Influenza dell'ambiente costiero e dello spray marino sulla pietra calcarea e sul marmot; Atti del 1º simposio internazionale, Bari, 7–10 giugno 1989,* ed. F. Zezza, 359–64. Brescia, Italy: Grafo Edizioni.

Redman, C. 1999. Cellulose sorbents: An evaluation of their working properties for use in wallpainting conservation. *The Conservator* 23: 68–76.

Reed, C. 2002. Soggy soil sours Egyptian sandstone. *Geotimes* 47 (9): 28–29.

Reedy, C. L. 2008. *Thin-Section Petrography of Stone and Ceramic Cultural Materials.* London: Archetype.

Ren, K. B., and D. A. Kagi. 1995. Upgrading the durability of mud bricks by impregnation. *Building and Environment* 30 (3): 433–40.

Ressl, C. 2007. Reconstruction of the Pegasus statue on top of the State Opera House in Vienna using photogrammetry and terrestrial and close-range laser scanning. In *Lasers in the Conservation of Artworks: LACONA VI, Vienna, Austria, Sept. 21–25, 2005; Proceedings,* ed. J. Nimmrichter, W. Kautek, and M. Schreiner, 535–42. Springer Proceedings in Physics 116. Berlin and New York: Springer.

Riecken, B., and H. R. Sasse. 1997. Durability of impregnation products for salt contaminated natural stone. In *Structural Studies, Repairs, and Maintenance of Historical Buildings (STREMAH): Proceedings of the Fifth International Conference, San Sebastián, Spain 1997,* ed. S. Sánchez-Beitia and C. A. Brebbia, 203–12. International Series on Advances in Architecture 3. Southampton, UK, and Boston: Computational Mechanics Publications.

Rijniers, L. A., H. P. Huinink, L. Pel, and K. Kopinga. 2005. Experimental evidence of crystallization pressure inside porous media. *Physical Review Letters* 94 (7): 075503-1–075503-4.

Rijniers, L. A., P. C. M. M. Magusin, H. P. Huinink, L. Pel, and K. Kopinga. 2004. Sodium NMR relaxation in porous materials. *Journal of Magnetic Resonance* 167 (1): 25–30.

RILEM (Réunion Internationale des Laboratoires et Experts des Matériaux, systèmes de construction et ouvrages) Commission 25-PEM (Protection et érosion des monuments). 1980. Essais recommandés pour mesurer l'altération des pierres et évaluer l'efficacité des méthodes de traitement = Recommended tests to measure the deterioration of stone and to assess the effectiveness of treatment methods. *Matériaux et constructions = Materials and Structures* 13 (3): 175–253.

RILEM (Réunion Internationale des Laboratoires et Experts des Matériaux, systèmes de construction et ouvrages) Commission 25-PEM (Protection et érosion des monuments). 1978. Essais recommandés pour mesurer l'altération des pierres et évaluer l'efficacité des méthodes de traitement = Recommended tests to measure the deterioration of stone and to assess the effectiveness of treatment methods. In *Altération et protection des monuments en pierre: Colloque international, Paris, du 5 au 9 juin 1978 = Deterioration and Protection of Stone Monuments: International symposium, Paris, June 5–9, 1978.* UNESCO and RILEM International Association of Testing and Research Laboratories for Materials and Structures 8. Paris: Centre d'études du batîment et des travaux publics.

Robertson, E. C. 1982. Physical properties of building stone. In *Conservation of Historic Stone Buildings and Monuments: Report of the Committee on Conservation of Historic Stone Buildings and Monuments, National Materials Advisory Board, Commission on Engineering and Technical Systems, National Research Council,* 62–86. Washington, DC: National Academy Press.

Roby, T. C. 1996. In situ assessment of surface consolidation and protection treatments of marble monuments in Rome of the 1980s, with particular reference to two treatments with Paraloid B72. In *Proceedings of the 8th International Congress on Deterioration and Conservation of Stone, Berlin, 30 Sept.–4 Oct. 1996,* ed. J. Riederer, 1015–28. Berlin: Möller Druck und Verlag.

Rockwell, C., ed. 1994. *1994 International Directory of Training in Conservation of Cultural Heritage.* 5th ed. Rome: ICCROM; Marina del Rey, CA: Getty Conservation Institute.

Rodríguez-Navarro, C., and E. Doehne. 1999. Salt weathering: Influence of evaporation rate, supersaturation and crystallization pattern. *Earth Surface Processes and Landforms* 24 (2–3): 191–209.

Rodríguez-Navarro, C., and E. Sebastian. 1996. Role of particulate matter from vehicle exhaust on porous building stones (limestone) sulfation. *Science of the Total Environment* 187 (2): 79–91.

Rodríguez-Navarro, C., E. Doehne, and E. Sebastian. 1999. Origins of honeycomb weathering: The role of salts and wind. *Bulletin of the Geological Society of America* 111 (8): 1250–55.

Rodríguez-Navarro, C., L. L. Hernandez, and E. Sebastian. 2006. New developments for preventing salt damage to porous ornamental materials through the use of crystallization inhibitors. In *Proceedings of the ARCCHIP Workshop ARIADNE 13: Problems of Salts in Masonry; "SALTeXPERT" (with the Getty Conservation Institute), November 27–December 1, 2002,* ed. S. Simon and M. Drdácký, vol. 5, 329–40. European Research on Cultural Heritage: State-of-the-Art Studies. Prague: Institute for Theoretical and Applied Mechanics, Academy of Sciences of the Czech Republic.

Rodríguez-Navarro, C., K. Elert, E. Sebastián, R. M. Esbert, C. M. Grossi, A. Rojo, F. J. Alonso, M. Montoto, and J. Ordaz. 2003. Laser cleaning of stone materials: An overview of current research. *Reviews in Conservation* (4): 65–82.

Rodríguez-Navarro, C., E. Sebastian, E. Doehne, and W. S. Ginell. 1998. The role of sepiolite-palygorskite in the decay of ancient Egyptian limestone sculptures. *Clays and Clay Minerals* 46 (4): 414–22.

Roels, S., J. Carmeliet, H. Hens, O. Adan, H. Brocken, R. Cerny, Z. Pavlik, C. Hall, K. Kumaran, L. Pel, and R. Plagge. 2004. Interlaboratory comparison of hygric properties of porous building materials. *Journal of Thermal Envelope and Building Science* 27 (4): 307–25.

Rosato, V. G. 2008. Lichens on religious buildings in La Plata and surrounding area (Buenos Aires province, Argentina). In *Proceedings of the 11th International Congress on Deterioration and Conservation of Stone, 15–20 September 2008, Toruń, Poland,* ed. J. W. Lukaszewicz and P. Niemcewicz, 243–49. Toruń, Poland: Nicolaus Copernicus University.

Rosenfeld, A. 1988. *Rock Art Conservation in Australia.* 2nd ed. Special Australian Heritage Publication Series 2. Canberra: Australian Govt. Pub. Service.

Rosling, H. 2009. Gapminder. Available at http://www.gapminder.org.

Rosvall, J., and B. Lagerqvist. 1993. A standardized monitoring system in cultural conservation: A tentative model. In *Conservation of Architectural Surfaces: Stones and Wall Covering,* ed. G. Biscontin and L. Graziano, 69–76. Venice: Il Cardo.

Rousset, B., G. Stéfania, J. James, and B. Pozzi. 2005. Injection grouts for molasse sandstones: Preliminary assessments. In *International RILEM Workshop "Repair Mortars for Historic Masonry," Delft University of Technology, The Netherlands, 26–28 January 2005: Proceedings.* http://www.conservation-science.ch/files/rousset_gentile_james_pozzi_2005_delft_injectiongroutsformolassesandstones preliminaryassessments.pdf.

Rowe, S., and C. Rozeik. 2008. The uses of cyclododecane in conservation. *Reviews in Conservation* 9: 17–31.

Rozenbaum, O., L. Barbanson, F. Muller, and A. Bruand. 2008. Significance of a combined approach for replacement stones in the heritage buildings' conservation frame. *Comptes Rendus—Geoscience* 340 (6): 345–55.

Ruedrich, J., T. Weiss, and S. Siegesmund. 2002. Thermal behaviour of weathered and consolidated marbles. In *Natural Stone, Weathering Phenomena, Conservation Strategies and Case Studies,* ed. S. Siegesmund, T. Weiss, and A. Vollbrecht, 255–71. Geological Society Special Publications 205. London: Geological Society of London.

Ruiz-Agudo, E., J. D. Martín-Ramos, and C. Rodríguez-Navarro. 2007. Mechanism and kinetics of dehydration of epsomite crystals formed in the presence of organic additives. *Journal of Physical Chemistry B* 111 (1): 41–52.

Ruiz-Agudo, E., C. V. Putnis, and C. Rodríguez-Navarro. 2008. Interaction between epsomite crystals and organic additives. *Crystal Growth and Design* 8 (8): 2665–73.

Ruiz de Argandoña, V. G., L. Calleja, Á. Rodríguez-Rey, L. M. Suárez del Río, and C. Celorio. 2009. X-ray computed tomography study of the influence of consolidants on the hydric properties of sandstones for stone conservation studies. *Engineering Geology* 103 (3–4): 69–75.

Saarela, M., H.-L. Alakomi, M.-L. Suihko, L. Maunuksela, L. Raaska, and T. Mattila-Sandholm. 2004. Heterotrophic microorganisms in air and biofilm samples from Roman catacombs, with special emphasis on actinobacteria and fungi. *International Biodeterioration and Biodegradation* 54 (1): 27–37.

Sabbioni, C., M. Cassar, P. Brimblecombe, J. Tilblad, R. Kozlowski, M. Drdacky, C. Saiz-Jimenez, T. Grontoft, I. Wainwrigh, and X. Arino. 2006. Global climate change impact on built heritage and cultural landscapes. In *Heritage, Weathering and Conservation: Proceedings of the International Conference on Heritage, Weathering and Conservation (HWC-2006), 21–24 June 2006, Madrid,* ed. R. Fort, M. Alvarez de Buergo, M. Gomez-Heras, and C. Vazquez-Calvo, 395–401. Balkema Proceedings and Monographs in Engineering, Water and Earth Sciences. London and New York: Taylor & Francis.

Sadat-Shojai, M., and A. Ershad-Langroudi. 2009. Polymeric coatings for protection of historic monuments: Opportunities and challenges. *Journal of Applied Polymer Science* 112 (4): 2535–51.

Sagai, M., A. Furuyama, and T. Ichinose. 1996. Biological effects of diesel exhaust particles (DEP), III: Pathogenesis of asthma like symptoms in mice. *Free Radical Biology and Medicine* 21 (2): 199–209.

Saidov, T., and L. Pel. 2008. The formation of meta stable sodium sulfate heptahydrate during drying as measured by NMR. In *Proceedings of the 11th International Congress on Deterioration and Conservation of Stone, 15–20 September 2008, Toruń, Poland,* ed. J. W. Lukaszewicz and P. Niemcewicz, 251–56. Toruń, Poland: Nicolaus Copernicus University.

Saiz-Jimenez, C. 1995. Deposition of anthropogenic compounds on monuments and their effect on airborne microorganisms. *Aerobiologia* 11 (3): 161–75.

Saiz-Jimenez, C., ed. 2004. *Air Pollution and Cultural Heritage: Proceedings of the International Workshop on Air Pollution and Cultural Heritage, 1–3 December 2003, Seville, Spain.* Leiden and New York: Balkema.

Saiz-Jimenez, C. and L. Laiz. 2000. Occurrence of halotolerant/halophilic bacterial communities in deteriorated monuments. In *International Biodeterioration & Biodegradation* 46 (4): 319–26.

Salazar-Hernández, C., R. Zárraga, S. Alonso, S. Sugita, S. Calixto, and J. Cervantes. 2009. Effect of solvent type on polycondensation of TEOS catalyzed by DBTL as used for stone consolidation. *Journal of Sol-Gel Science and Technology* 49 (3): 301–10.

Sanchez-Moral, S., L. Luque, S. Cuezva, V. Soler, D. Benavente, L. Laiz, J. M. Gonzalez, and C. Saiz-Jimenez. 2005. Deterioration of building materials in Roman catacombs: The influence of visitors. *Science of the Total Environment* 349 (1–3): 260–76.

Santoro, E. D., and R. J. Koestler. 1991. A methodology for biodeterioration testing of polymers and resins. *International Biodeterioration* 28 (1–4): 81–90.

Sarró, M. I., A. M. García, V. M. Rivalta, D. A. Moreno, and I. Arroyo. 2006. Biodeterioration of the Lions Fountain at the Alhambra Palace, Granada (Spain). *Building and Environment* 41 (12): 1811–20.

Sasse, H. R., and D. Honsinger. 1991. A new chemical and engineering approach for development and optimization of stone protecting materials. In *Science, Technology, and European Cultural Heritage: Proceedings of the European Symposium, Bologna, Italy, 13–16 June 1989,* ed. N. S. Baer, C. Sabbioni, and A. I. Sors, 649–52. Oxford and Boston: Published for the Commission of the European Communities by Butterworth-Heinemann Publishers.

Sasse, H. R., and B. Riecken. 1993. "VENUS": A new concept in laboratory simulation of aggressive climatic conditions. In *Conservation of Stone and Other Materials: Proceedings of the International RILEM/UNESCO Congress "Conservation of Stone and Other Materials: Research – Industry – Media," UNESCO Headquarters, Paris, June 29–July 1, 1993,* ed. M.-J. Thiel, 801–7. RILEM Proceedings 21. London and New York: E & FN Spon.

Sasse, H. R., and R. Snethlage. 1996. Evaluation of stone consolidation treatments. *Science and Technology for Cultural Heritage* 5 (1): 85–92.

Sasse, H. R., and R. Snethlage. 1997. Methods for the evaluation of stone conservation treatments. In *Saving Our Architectural Heritage: The Conservation of Historic Stone Structures; Report of the Dahlem Workshop on Saving Our Architectural Heritage, The Conservation of Historic Stone Structures, Berlin, March 3–8, 1996*, ed. N. S. Baer and R. Snethlage, 223–44. Dahlem Workshop Reports. Chichester and New York: John Wiley & Sons.

Sawdy, A. 1995. An examination of the kinetics of deliquescence and recrystallisation of some soluble salts found in wall paintings. Master's thesis, University of London, Courtauld Institute of Art, Conservation of Wall Painting Department.

Sawdy, A. M. 2002. The kinetics of salt weathering of porous materials: Stone monuments and wall paintings. PhD diss., University of London.

Sawdy, A., and A. Heritage. 2007. Evaluating the influence of mixture composition on the kinetics of salt damage in wall paintings using time-lapse video imaging with direct data annotation. *Environmental Geology* 52 (2): 303–15.

Sawdy, A., and C. Price. 2005. Salt damage at Cleeve Abbey, England, Part I: A comparison of theoretical predictions and practical observations. *Journal of Cultural Heritage* 6 (2): 125–35.

Sawdy, A., A. Heritage, and L. Pel. 2008. A review of salt transport in porous media: Assessment methods and salt reduction treatments. In *Salt Weathering on Buildings and Stone Sculptures, 22–24 October 2008, The National Museum Copenhagen, Denmark* [Proceedings from the international conference], ed. J. S. Albertsen, 1–27. Lyngby: Technical University of Denmark, Department of Civil Engineering.

Schaffer, R. J. 1932. *The Weathering of Natural Building Stones*. Department of Scientific and Industrial Research, Building Research Board, Special Report 18. London: H.M. Stationery Office [printed by Harrison and Sons].

Scheerer, S., O. Ortega-Morales, and C. Gaylarde. 2009. Microbial deterioration of stone monuments: An updated overview. *Advances in Applied Microbiology* 66: 97–139.

Scheffler, M. J. 2001. Thin-stone veneer building facades: Evolution and preservation. *APT Bulletin* 32 (1): 27–34.

Scherer, G. W. 1999. Crystallization in pores. *Cement and Concrete Research* 29 (8): 1347–58.

Scherer, G. W. 2000. Stress from crystallization of salt in pores. In *Proceedings of the 9th International Congress on Deterioration and Conservation of Stone, Venice, June 19–24, 2000,* ed. V. Fassina, vol. 1, 187–94. Amsterdam and New York: Elsevier.

Scherer, G. W. 2004. Stress from crystallization of salt. *Cement and Concrete Research* 34 (9): 1613–24.

Scherer, G. W. 2006. Internal stress and cracking in stone and masonry. In *Measuring, Monitoring and Modeling Concrete Properties: An International Symposium Dedicated to Professor Surendra P. Shah, Northwestern University, U.S.A.*, ed. M. S. Konsta-Gdoutos, 633–41. Dordrecht, Netherlands: Springer.

Scherer, G., and I. Jiménez-González. 2008. Swelling clays and salt crystallization: The damage mechanisms and the role of consolidants. In *Stone Consolidation in Cultural Heritage: Research and Practice; Proceedings of the International Symposium, Lisbon, 6–7 May 2008,* ed. J. Delgado Rodrigues and J. M. Mimoso, 29–40. Lisbon: LNEC (Laboratório Nacional de Engenharia Civil).

Scherer, G. W., and J. J. Valenza II. 2005. Mechanisms of frost damage. In *Materials Science of Concrete VII,* ed. F. Young and J. Skalny, vol. 7, 209–46. Westerville, OH: American Ceramic Society.

Scherer, G. W., and G. S. Wheeler. 2009. Silicate consolidants for stone. *Key Engineering Materials* 391: 1–25.

Scherer, G. W., R. Flatt, and G. Wheeler. 2001. Materials science research for the conservation of sculpture and monuments. *MRS Bulletin* 26 (1): 44–50.

Schiavon, N. 1992. Decay mechanisms of oolithic limestones in an urban environment: King's College Chapel, Cambridge and St. Luke's Church, London. In *Stone Cleaning and the Nature, Soiling and Decay Mechanisms of Stone: Proceedings of the International Conference Held in Edinburgh, UK, 14–16 April 1992,* ed. R. G. M. Webster, 258–67. London: Donhead Publishing.

Schiavon, N. 2002. Biodeterioration of calcareous and granitic building stones in urban environments. In *Natural Stone, Weathering Phenomena, Conservation Strategies and Case Studies,* ed. S. Siegesmund, T. Weiss, and A. Vollbrecht, 195–205. Geological Society Special Publications 205. London: Geological Society of London.

Schiavon, N. 2007. Kaolinisation of granite in an urban environment. *Environmental Geology* 52 (2): 399–407.

Schiavon, N., G. Chiavari, and D. Fabbri. 2004. Soiling of limestone in an urban environment characterized by heavy vehicular exhaust emissions. *Environmental Geology* 46 (3–4): 448–55.

Schmidt, H. 1999. Historische bauwerksabdichtungen: Traditionelle und neuzeitliche maßnahmen zum schutz gegen bodenfeuchtigkeit und zur trockenlegung feuchter wände [Historical waterproofing: Traditional and modern methods of moisture protection and wall drainage]. *Bautechnik* 76 (7): 581–98.

Schnabel, L. 1992. Evaluation of the barium hydroxide-urea consolidation method. In *Proceedings of the 7th International Congress on Deterioration and Conservation of Stone, Lisbon, 15–18 June 1992,* ed. J. Delgado Rodrigues,

F. Henriques, and F. Telmo Jeremias, 1063–72. Lisbon: Laboratório Nacional de Engenharia Civil.

Schoonbrood, J. W. M. 1993. Low pressure application technique for stone preservatives. In *Conservation of Stone and Other Materials: Proceedings of the International RILEM/UNESCO Congress "Conservation of Stone and Other Materials: Research—Industry—Media," UNESCO Headquarters, Paris, June 29–July 1, 1993,* ed. M.-J. Thiel, 512–18. RILEM Proceedings 21. London and New York: E & FN Spon.

Schreiber, K. V., and T. C. Meierding. 1999. Spatial patterns and causes of marble tombstone weathering in western Pennsylvania. *Physical Geography* 20 (3): 173–88.

Schvoerer, M., ed. 1999. *Archéomatériaux: Marbres et autres roches; Actes de la IVème Conférence internationale de l'Association pour l'étude des marbres et autres roches utilisés dans le passé, ASMOSIA IV, Bordeaux-Talence, 9–13 octobre 1995.* ASMOSIA no. 4. Talence, France: CRPAA, PUB. Centre de recherche en physique appliquée à l'archéologie; Presses Universitaires de Bordeaux.

Searle, D. E., and D. J. Mitchell. 2004. Stone weathering and urban particulate in the UK. In *Stone Deterioration in Polluted Urban Environments,* ed. D. J. Mitchell, and D. E. Searle, 1–22. Land Reconstruction and Management Series 3. Enfield, NH: Science Publishers.

Searle, D. E., and D. J. Mitchell. 2006. The effect of coal and diesel particulates on the weathering loss of Portland limestone in an urban environment. *Science of the Total Environment* 370 (1): 207–23.

Searle, D. E., and D. J. Mitchell. 2008. A methodology for the semi-quantitative analysis of soiling patterns on building facades in Bath, U.K., using digital image processing. In *Proceedings of the 11th International Congress on Deterioration and Conservation of Stone, 15–20 September 2008, Toruń, Poland,* ed. J. W. Lukaszewicz and P. Niemcewicz, 497–504. Toruń, Poland: Nicolaus Copernicus University.

Sedlbauer, K., and H. M. Künzel. 2000. Frost damage of masonry walls: A hygrothermal analysis by computer simulations. *Journal of Thermal Envelope and Building Science* 23 (January): 277–81. http://www.hoki.ibp.fhg.de/ibp/publikationen/fachzeitschriften/frost_damage_1999_190602.pdf.

Selwitz, C. M. 1991. The use of epoxy resins for stone consolidation. In *Materials Issues in Art and Archaeology II Symposium, Held April 17–21, 1990, San Francisco,* ed. P. B. Vandiver, J. R. Druzik, and G. S. Wheeler, 181–91. Materials Research Society Symposium Proceedings 185. Pittsburgh: Materials Research Society.

Selwitz, C. 1992a. *Epoxy Resins in Stone Conservation.* Research in Conservation 7. Marina del Rey, CA: Getty Conservation Institute. http://www.getty.edu/conservation/publications/pdf_publications/epoxyresins.pdf.

Selwitz, C. 1992b. The use of epoxy resins in field projects for stone stabilization. In *Materials Issues in Art and Archaeology III Symposium, Held April 27–May 1, 1992, San Francisco,* ed. P. B. Vandiver, J. R. Druzik, G. S. Wheeler, and I. C. Freestone, 925–34. Materials Research Society Symposium Proceedings 267. Pittsburgh: Materials Research Society.

Selwitz, C., and E. Doehne. 2002. The evaluation of crystallization modifiers for controlling salt damage to limestone. *Journal of Cultural Heritage* 3 (3): 205–16.

Shahidzadeh, N., S. Rafai, D. Bonn, and G. Wegdam. 2008. Salt crystallization during evaporation: Impact of interfacial properties. In *Proceedings of the 11th International Congress on Deterioration and Conservation of Stone, 15–20 September 2008, Toruń, Poland,* ed. J. W. Lukaszewicz and P. Niemcewicz, 257–65. Toruń, Poland: Nicolaus Copernicus University.

Sharma, R. K., and H. O. Gupta. 1993. Dust pollution at the Taj Mahal: A case study. In *Conservation of Stone and Other Materials: Proceedings of the International RILEM/UNESCO Congress "Conservation of Stone and Other Materials: Research—Industry—Media," held at UNESCO Headquarters, Paris, June 29–July 1, 1993,* ed. M.-J. Thiel, 11–18. RILEM Proceedings 21. London and New York: E & FN Spon.

Siano, S., M. Giamello, L. Bartoli, A. Mencaglia, V. Parfenov, and R. Salimbeni. 2008. Laser cleaning of stone by different laser pulse duration and wavelength. *Laser Physics* 18 (1): 27–36.

Siedel, H. 2008. Salt-induced alveolar weathering of rhyolite tuff on a building: Causes and processes. In *Salt Weathering on Buildings and Stone Sculptures, 22–24 October 2008, The National Museum Copenhagen, Denmark* [Proceedings from the international conference], ed. J. S. Albertsen, 79–88. Lyngby: Technical University of Denmark, Department of Civil Engineering. http://www.tu-dresden.de/biw/geotechnik/geologie/publikationen/download/Alveolar_Copenhagen_2008.pdf.

Siedel, H., K. Neumeister, and R. J. G. Sobott. 2003. Laser cleaning as a part of the restoration process: Removal of aged oil paints from a Renaissance sandstone portal in Dresden, Germany. *Journal of Cultural Heritage* 4 (Suppl. 1: Lasers in the Conservation of Artworks—LACONA IV): 11–16.

Siedel, H., E. von Plehwe-Leisen, and H. Leisen. 2008. Salt load and deterioration of sandstone at the temple of Angkor Wat, Cambodia. In *Proceedings of the 11th International Congress on Deterioration and Conservation of Stone, 15–20 September 2008, Toruń, Poland,* ed. J. W. Lukaszewicz and P. Niemcewicz, 267–74. Toruń, Poland: Nicolaus Copernicus University.

Siegesmund, S., A. Koch, and J. Ruedrich. 2007. Marble as building veneer: Weathering dynamics and bowing potentials = Fassadenplatten aus Marmor: Verwitterungsdynamik und Verbiegungspotentiale. *Zeitschrift der Deutschen Gesellschaft für Geowissenschaften* 158 (3): 631–47.

Siegesmund, S., J. Ruedrich, and A. Koch. 2008. Marble bowing: Comparative studies of three different public building facades. *Environmental Geology* 56 (3–4): 473–94.

Siegesmund, S., R. Snethlage, and J. Ruedrich. 2008. Monument futures: Climate change, air pollution, decay and conservation; The Wolf-Dieter Grimm-volume. *Environmental Geology* 56 (3–4): 451–53.

Siegesmund, S., T. Weiss, and S. Rüdrich. 2004. Ultrasonic wave velocities as a diagnostic tool for the quality assessment of marbles. In *Influence of the Environment and Defense of the Territory on Recovery of Cultural Heritage: Proceedings of the 6th International Symposium on the Conservation of Monuments in the Mediterranean Basin, Lisbon, Portugal, April 7–10, 2004,* ed. L. Aires-Barros, F. Zezza, A. Dionísio, and M. Rodrigues, 342–46. Lisbon: International Group for the Conservation of Monuments in the Mediterranean Basin.

Siegesmund, S., A. Török, A. Hüpers, C. Müller, and W. Klemm. 2007. Mineralogical, geochemical and microfabric evidences of gypsum crusts: A case study from Budapest. *Environmental Geology* 52 (2): 385–97.

Sikiotis, D., and P. Kirkitsos. 1995. The adverse effects of nitrates on stone monuments. *Science of the Total Environment* 171 (1–3): 173–82.

Sikka, S., C. Selwitz, E. Doehne, G. Chiari, and H. Khanjian. 2008. Qualitative and quantitative methods of detection of calcium oxalate deposits on treated limestone and marble. In *Stone Consolidation in Cultural Heritage: Research and Practice; Proceedings of the International Symposium, Lisbon, 6–7 May 2008,* ed. J. Delgado Rodrigues and J. M. Mimoso, 445–54. Lisbon: LNEC (Laboratório Nacional de Engenharia Civil).

Silva, B., N. Aira, A. Martínez-Cortizas, and B. Prieto. 2009. Chemical composition and origin of black patinas on granite. *Science of the Total Environment* 408 (1): 130–37.

Silva, Z. S. G., and J. A. R. Simão. 2009. The role of salt fog on alteration of dimension stone. *Construction and Building Materials* 23 (11): 3321–27.

Simão, J., E. Ruiz-Agudo, and C. Rodríguez-Navarro. 2006. Effects of particulate matter from gasoline and diesel vehicle exhaust emissions on silicate stones sulfation. *Atmospheric Environment* 40 (36): 6905–17.

Simionescu, B., M. Olaru, M. Aflori, E. C. Buruiana, and C. Cotofana. 2009. Silsesquioxane-based hybrid nanocomposites of polymethacrylate type with self-assembling properties. *Solid State Phenomena* 151: 17–23.

Simon, S., and E. Doehne. 2006a. Summary of discussion. In *Proceedings of the ARCCHIP Workshop ARIADNE 13: Problems of Salts in Masonry; "SALTeXPERT" (with the Getty Conservation Institute), November 27–December 1, 2002,* ed. S. Simon and M. Drdácký, vol. 5, 15–20. European Research on Cultural Heritage: State-of-the-Art Studies. Prague: Institute for Theoretical and Applied Mechanics, Academy of Sciences of the Czech Republic.

Simon, S., and E. Doehne. 2006b. Summary of proceedings. In *Proceedings of the ARCCHIP Workshop ARIADNE 13: Problems of Salts in Masonry; "SALTeXPERT" (with the Getty Conservation Institute), November 27–December 1, 2002*, ed. S. Simon and M. Drdácký, vol. 5, 9–13. European Research on Cultural Heritage: State-of-the-Art Studies. Prague: Institute for Theoretical and Applied Mechanics, Academy of Sciences of the Czech Republic.

Simon, S., and A.-M. Lind. 1999. Decay of limestone blocks in the block fields of Karnak Temple (Egypt): Non-destructive damage analysis and control of consolidation treatments. In *12th Triennial Meeting, ICOM Committee for Conservation, Lyon, 29 August–3 September 1999: Preprints*, ed. J. Bridgland, 743–49. London: James & James (Science Publishers) Ltd.

Simon, S., M. Shaer, and E. Kaiser. 2006. The salt-laden rock-carved tomb facades in Petra, Jordan: Scientific investigations. In *Proceedings of the ARCCHIP Workshop ARIADNE 13: Problems of Salts in Masonry; "SALTeXPERT" (with the Getty Conservation Institute), November 27–December 1, 2002*, ed. S. Simon and M. Drdácký, vol. 5, 341–49. European Research on Cultural Heritage: State-of-the-Art Studies. Prague: Institute for Theoretical and Applied Mechanics, Academy of Sciences of the Czech Republic.

Simon, S., V. Vergès-Belmin, A. Blanc, and R. Snethlage. 1994. The marble columns in the cloister of the Primatial Church St-Trophime in Arles, France: Ultrasonic velocity measurements as a tool for anamnesis and diagnosis. In *EUROCARE-EUROMARBLE EU 496: Proceedings of the Fourth Workshop, Arles, France, November 3–6, 1993*, ed. R. Snethlage and V. Vergès-Belmin, 11–17. Forschungsbericht (Bayerisches Landesamt für Denkmalpflege) 13. Munich: Bayerisches Landesamt für Denkmalpflege.

Skoulikidis, T. N., and N. Beloyannis. 1984. Inversion of marble sulfation-reconversion of gypsum films into calcite on the surfaces of monuments and statues. *Studies in Conservation* 29 (4): 197–204.

Slavid, I. O., and N. R. Weiss (inventors). 2001. Method for Protecting and Consolidating Calcareous Materials. U.S. Patent 6296905, filed October 28, 1999, and issued October 2, 2001.

Sleater, G. A. 1977. *Stone Preservatives: Methods of Laboratory Testing and Preliminary Performance Criteria*. NBS Technical Note 941. Washington, DC: U.S. Dept. of Commerce, National Bureau of Standards (Government Printing Office).

Smith, B. 2006. Rock art tourism in Southern Africa: Problems, possibilities, and poverty relief. In *Of the Past, for the Future: Integrating Archaeology and Conservation; Proceedings of the Conservation Theme at the 5th World Archaeological Congress, Washington, D.C., 22–26 June 2003*, ed. N. Agnew and J. Bridgland, 322–30. Getty Conservation Institute Symposium Proceedings Series. Los Angeles: Getty Conservation Institute.

Smith, B. J., M. Gómez-Heras, and S. McCabe. 2008. Understanding the decay of stone-built cultural heritage. *Progress in Physical Geography* 32 (4): 439–61.

Smith, B. J., and P. A. Warke, eds. 1996. *Processes of Urban Stone Decay: Proceedings of SWAPNET '95; Stone Weathering and Atmospheric Pollution Network Conference Held in Belfast, 19–20 May 1995*. London: Donhead Publishing.

Smith, B. J., M. Gómez-Heras, J. Meneely, S. McCabe, and H. A. Viles. 2008. High resolution monitoring of surface morphological change of building limestones in response to simulated salt weathering. In *Proceedings of the 11th International Congress on Deterioration and Conservation of Stone, 15–20 September 2008, Toruń, Poland*, ed. J. W. Lukaszewicz and P. Niemcewicz, 275–82. Toruń, Poland: Nicolaus Copernicus University.

Smith, B. J., P. A. Warke, J. P. McGreevy, and H. L. Kane. 2005. Salt-weathering simulations under hot desert conditions: Agents of enlightenment or perpetuators of preconceptions? *Geomorphology* 67 (1–2): 211–27.

Snethlage, R. 2002. Silicon organic compounds for natural stone conservation: New products and laboratory tests. In *I silicati nella conservazione: Indagini, esperienze e valutazioni per il consolidamento dei manufatti storici: Congresso internazionale, 13–15 febbraio, 2002, Villa Gualino, Villa Rey, Torino, Italy = Silicates in Conservative Treatments: Tests, Improvements and Evaluations of Consolidating Performance: International Congress, February 13–15, 2002, Villa Gualino, Villa Rey, Turin, Italy*, 27–36. Torino: Fondazione per le biotecnologie.

Snethlage, R. 2005. *Leitfaden Steinkonservierung: Planung von Untersuchungen und Massnahmen zur Erhaltung von Denkmälern aus Naturstein.* 2nd rev. ed. Stuttgart: Fraunhofer IRB Verlag.

Snethlage, R. 2008. *Leitfaden Steinkonservierung: Planung von Untersuchungen und Massnahmen zur Erhaltung von Denkmälern aus Naturstein.* 3rd ed. Stuttgart: IRB Verlag.

Snethlage, R., and E. Wendler. 1991. Surfactants and adherent silicon resins: New protective agents for natural stone. In *Materials Issues in Art and Archaeology II Symposium, Held April 17–21, 1990, San Francisco,* ed. P. B. Vandiver, J. R. Druzik, and G. S. Wheeler, 193–200. Materials Research Society Symposium Proceedings 185. Pittsburgh: Materials Research Society.

Snethlage, R., and E. Wendler. 1997. Moisture cycles and sandstone degradation. In *Saving Our Architectural Heritage: The Conservation of Historic Stone Structures; Report of the Dahlem Workshop on Saving Our Architectural Heritage, The Conservation of Historic Stone Structures, Berlin, March 3–8, 1996,* ed. N. S. Baer and R. Snethlage, 7–24. Dahlem Workshop Reports. Chichester and New York: John Wiley & Sons.

Snethlage, R., and E. Wendler. 2000. Chemical conservation of stone structures. In *Ullmann's Encyclopedia of Industrial Chemistry: 2000 Electronic Release,* ed. J. E. Bailey. New York: Wiley VCH. DOI: 10.1002/14356007.d06_d01.

Snethlage, R., E. Wendler, and L. Sattler. 1991. The application of laboratory processes and studies to real structures. In *Science, Technology, and European Cultural Heritage: Proceedings of the European Symposium, Bologna, Italy, 13–16 June 1989,* ed. N. S. Baer, C. Sabbioni, and A. I. Sors, 255–68. Oxford and Boston: Published for the Commission of the European Communities by Butterworth-Heinemann Publishers.

Snethlage, R., C. Gruber, V. Tucic, and W. Wendler. 2008. Transforming gypsum into calcium phosphate: A better way to preserve lime paint layers on natural stone? In *Stone Consolidation in Cultural Heritage: Research and Practice; Proceedings of the International Symposium: Lisbon, 6–7 May, 2008,* ed. J. Delgado Rodrigues and J. M. Mimoso, 1–14. Lisbon: LNEC (Laboratório Nacional de Engenharia Civil).

Solga, A., Z. Cerman, B. F. Striffler, M. Spaeth, and W. Barthlott. 2007. The dream of staying clean: Lotus and biomimetic surfaces. *Bioinspiration and Biomimetics* 2 (4): S126–34.

Sorlini, C., D. Falappi, and P. Sardi. 1991. Effects of mould growth on Serena stone samples untreated and treated with resin. In *Science, Technology, and European Cultural Heritage: Proceedings of the European Symposium, Bologna, Italy, 13–16 June 1989,* ed. N. S. Baer, C. Sabbioni, and A. I. Sors, 632–35. Oxford and Boston: Published for the Commission of the European Communities by Butterworth-Heinemann Publishers.

Sramek, J. 1993. Effects of hydrophobization on stone. In *Conservation of Architectural Surfaces: Stones and Wall Covering,* ed. G. Biscontin and L. Graziano, 169–74. Venice: Il Cardo.

St. Clair, L. L., and M. R. D. Seaward, eds. 2004. *Biodeterioration of Stone Surfaces: Lichens and Biofilms as Weathering Agents of Rocks and Cultural Heritage.* Dordrecht, Netherlands and London: Kluwer Academic Publishers.

Stancliffe, M., I. De Witte, and E. De Witte. 2005. St Paul's Cathedral: Poultice cleaning of the interior. *Journal of Architectural Conservation* 11 (3): 87–103.

Stefanis, N.-A., P. Theoulakis, and C. Pilinis. 2009. Dry deposition effect of marine aerosol to the building stone of the medieval city of Rhodes, Greece. *Building and Environment* 44 (2): 260–70.

Steiger, M. 2003. Salts and crusts. In *The Effects of Air Pollution on the Built Environment,* ed. P. Brimblecombe, 133–81. Air Pollution Reviews 2. London: Imperial College Press; Singapore and River Edge, NJ: World Scientific.

Steiger, M. 2005a. Crystal growth in porous materials II: Influence of crystal size on the crystallization pressure. *Journal of Crystal Growth* 282 (3–4): 470–81.

Steiger, M. 2005b. Salts in porous materials: Thermodynamics of phase transitions, modeling and preventive conservation. *Restoration of Buildings and Monuments: An International Journal = Bauinstandsetzen und Baudenkmalpflege: Eine internationale Zeitschrift* 11 (6): 419–32.

Steiger, M. 2006. Freezing of salt solutions in small pores. In *Measuring, Monitoring and Modeling Concrete Properties: An International Symposium Dedicated to Professor Surendra P. Shah, Northwestern University, U.S.A.*, ed. M. S. Konsta-Gdoutos, 661–68. Dordrecht, Netherlands: Springer.

Steiger, M., J. Kiekbusch, and A. Nicolai. 2008. An improved model incorporating Pitzer's equations for calculation of thermodynamic properties of pore solutions implemented into an efficient program code. *Construction and Building Materials* 22 (8): 1841–50.

Steiger, M., and S. Siegesmund. 2007. Special issue on salt decay, *Environmental Geology* 52 (2): 185–86.

Steiger, M., and A. Zeunert. 1996. Crystallization properties of salt mixtures: Comparison of experimental results and model calculations. In *Proceedings of the 8th International Congress on Deterioration and Conservation of Stone, Berlin, 30 Sept.–4 Oct. 1996,* ed. J. Riederer, 535–44. Berlin: Möller Druck und Verlag.

Stein, R., J. Kimmel, M. Marincola, and F. Klemm. 2000. Observations on cyclododecane as a temporary consolidant for stone. *Journal of the American Institute for Conservation* 39: 355–69.

Stepien, P., R. Kozlowski, and M. Tokarz. 1993. Gypstop—colloidal silica for protective coating of porous building materials: Practical experience at the Wawel Castle, Cracow, Poland. In *Structural Repair and Maintenance of Historical Buildings III,* ed. C. A. Brebbia and R. J. B. Frewer, 304–10. Southampton and Boston: Computational Mechanics Publications.

Stieber, M. C. 1995. Book review: The Art of Stoneworking: A Reference Guide. *American Journal of Archaeology* 99 (3): 536–37.

Stone Federation of Great Britain. 1991. *Natural Stone Glossary*. London: Linked Advertising & Marketing.

Stulik, D., D. Miller, H. Khanjian, N. Khandekar, R. Wolbers, J. Carlson, W. Christian Petersen, and V. Dorge, eds. 2004. *Solvent Gels for the Cleaning of Works of Art: The Residue Question*. Research in Conservation. Los Angeles: Getty Conservation Institute.

Sumner, P. D., D. W. Hedding, and K. I. Meiklejohn. 2007. Rock surfaces temperatures in southern Namibia and implications for thermally-driven physical weathering. *Zeitschrift für Geomorphologie* 51 (Suppl. 1): 133–47.

Svahn, H. 2006. *Non-Destructive Field Tests in Stone Conservation: Field and Laboratory Tests: Final Report for the Research and Development Project*. Rapport från Riksantikvarieämbetet 2006:4. Stockholm: Riksantikvarieämbetet. http://www.raa.se/publicerat/9172094354.pdf.

Swartz, B. K., Jr., and J. P. Hale. 2000. Establishment of a global archive of prehistoric rock art photographs. *Tracce: Online Rock Art Bulletin* (12). http://www.rupestre.net/tracce/.

Taboada Téllez, F. 2007. Diagnóstico de conservación del Sitio Juan Miserandino, Municipio de Roboré, Depto. de Santa Cruz. *Boletín* (Sociedad de Investigación del Arte Rupestre de Bolivia) 21: 38–45.

Taruvinga, P. 2003. Salvaging vandalised rock art at Domboshava National Monument in North-eastern Zimbabwe. In *Heritage at Risk: ICOMOS World Report 2001/2002 on Monuments and Sites in Danger = Patrimoine en péril: ICOMOS rapport mondial 2001/2002 sur des monuments et des sites en péril = Patrimonio en peligro: ICOMOS informe mundial 2001/2002 sobre monumentos y sitios en peligro,* ed. D. Bumbaru, 217–19. International Council on Monuments and Sites. Munich: K.G. Saur. http://www.international.icomos.org/risk/2001/zimb2001.htm#.

Tavukçuoglu, A., A. Düzgüneş, E. N. Caner-Saltik, and S. Demirci. 2005. Use of IR thermography for the assessment of surface-water drainage problems in a historical building, AgzIkarahan (Aksaray), Turkey. *NDT & E International* 38 (5): 402–10.

Teutonico, J. M. 2004. Conclusions and recommendations of the Colloquium "Reburial of Archaeological Sites," Santa Fe, New Mexico, 17–21 March 2003. *Conservation and Management of Archaeological Sites* 6 (3–4): 395–99.

Teutonico, J. M., A. E. Charola, E. De Witte, G. Grassegger, R. J. Koestler, M. Laurenzi Tabasso, H. R. Sasse, and R. Snethlage. 1997. Group report: How can we ensure the responsible and effective use of treatments (cleaning, consolidation, protection)? In *Saving Our Architectural Heritage: The Conservation of Historic Stone Structures; Report of the Dahlem Workshop on Saving Our Architectural Heritage, The Conservation of Historic Stone Structures, Berlin, March 3–8, 1996*, ed. N. S. Baer and R. Snethlage, 293–313. Dahlem Workshop Reports. Chichester and New York: John Wiley & Sons.

Thakre, R., A. L. Aggorwal, and P. Khanna. 1997. Regional air environment management strategies for the protection of Taj Mahal and other monuments in India. In *Air Quality Management at Urban, Regional and Global Scales: Proceedings of the 10th Regional IUAPPA Conference, Gümüssuyu, Istanbul, Turkey, September 23–26, 1997*, ed. S. Incecik, 616–23. Environmental Research Forum 7–8. Zurich and Enfield, NH: Trans Tech Publications.

Theoulakis, P., N. A. Stefanis, I. Karatasios, V. Kilikoglou, and C. Pilinis. 2008. Performance evaluation of surface coatings against the degradation of stone monuments caused by the deposition of airborne particulate matter. In *Proceedings of the 11th International Congress on Deterioration and Conservation of Stone, 15–20 September 2008, Toruń, Poland*, ed. J. W. Lukaszewicz and P. Niemcewicz, 1071–76. Toruń, Poland: Nicolaus Copernicus University.

Thorn, A., and A. Piper. 1996. The Isle of the Dead: An integrated approach to the management and natural protection of an archaeological site. In *Archaeological Conservation and Its Consequences: Preprints of the Contributions to the Copenhagen Congress, 26–30 August 1996*, ed. A. Roy and P. Smith, 149–52. London: International Institute for Conservation of Historic and Artistic Works.

Thornbush, M. J., and H. A. Viles. 2008. Photographic monitoring of soiling and decay of roadside walls in central Oxford, England. *Environmental Geology* 56 (3–4): 777–87.

Tiano, P. 1994. Biodeterioration of stone monuments: A critical review. In *Recent Advances in Biodeterioration and Biodegradation*, ed. K. L. Garg, N. Garg, and K. G. Mukerji, vol. 1, 301–21. Calcutta: Naya Prokash.

Tiano, P. 1995. Stone reinforcement by calcite crystals precipitation induced by organic matrix macromolecules. *Studies in Conservation* 40 (3): 171–76.

Tiano, P. 2004. Innovative treatments for stone conservation. *Corrosion Reviews* 22 (5–6): 365–80.

Tiano, P. 2008. Biocalcification: The context for bioremediation. In *Heritage Microbiology and Science: Microbes, Monuments and Maritime Materials*, ed. E. May, M. Jones, and J. Mitchell, 62–75. Royal Society of Chemistry (Great Britain) Special Publication 315. Cambridge: Royal Society of Chemistry. http://www.rsc.org/ebooks/2008/BK9780854041411/BK9780854041411-00062.pdf.

Tiano, P., and C. Pardini, eds. 2008. *In Situ Monitoring of Monumental Surfaces: Proceedings of the International Workshop SMW08, 27–29 October 2008, Florence, Italy*. Florence: Edifir.

Tiano, P., L. Addadi, and S. Weiner. 1992. Stone reinforcement by induction of calcite crystals using organic matrix macromolecules: Feasibility study. In *Proceedings of the 7th International Congress on Deterioration and Conservation of Stone, Lisbon, 15–18 June 1992*, ed. J. Delgado Rodrigues, F. Henriques, and F. Telmo Jeremias, 1317–26. Lisbon: Laboratório Nacional de Engenharia Civil.

Toniolo, L., F. Cappitelli, C. Sorlini, and D. Gulotta. 2008. The biological approach for the removal of black crusts from stone surface of historical monuments. In *Proceedings of the 11th International Congress on Deterioration and Conservation of Stone, 15–20 September 2008, Toruń, Poland*, ed. J. W. Lukaszewicz and P. Niemcewicz, 1077–84. Toruń, Poland: Nicolaus Copernicus University.

Toniolo, L., C. M. Zerbi, and R. Bugini. 2009. Black layers on historical architecture. *Environmental Science and Pollution Research* 16 (2): 218–26.

Toniolo, L., C. Colombo, M. Realini, A. Peraio, and M. Positano. 2001. Evaluation of barium hydroxide treatment efficacy on a dolomitic marble. *Annali di Chimica* 91 (11–12): 813–21.

Török, Á. 2008. The application of silica acid ester under vacuum conditions for in situ consolidation of porous limestone monument: A case study from Hungary. In *Proceedings of the 11th International Congress on Deterioration and Conservation of Stone, 15–20 September 2008, Toruń, Poland,* ed. J. W. Lukaszewicz and P. Niemcewicz, 1085–91. Toruń, Poland: Nicolaus Copernicus University.

Torraca, G. 1999. The scientist in conservation. *Conservation: The GCI Newsletter* 14 (3). http://www.getty.edu/conservation/publications/newsletters/14_3/feature1_3.html.

Torraca, G. 2009. *Lectures on Materials Science for Architectural Conservation.* Los Angeles: Getty Conservation Institute. http://www.getty.edu/conservation/publications/pdf_publications/torraca.pdf.

Torrisi, A. 2008. XPS study of five fluorinated compounds deposited on calcarenite stone, Part II: Aged samples. *Applied Surface Science* 254 (22): 7127–36.

Tratebas, A. M., N. V. Cerveny, and R. I. Dorn. 2004. The effects of fire on rock art: Microscopic evidence reveals the importance of weathering rinds. *Physical Geography* 25 (4): 313–33.

Trentelman, K. 2005. Training and education in conservation science. *Conservation: The GCI Newsletter* 20 (2): 17–20. http://www.getty.edu/conservation/publications/newsletters/20_2/news_in_cons1.html.

Trinks, I., M. Díaz-Andreu, R. Hobbs, and K. E. Sharpe. 2005. Digital rock art recording: Visualising petroglyphs using 3D laser scanner data. *Rock Art Research: The Journal of the Australian Rock Art Research Association (AURA)* 22 (2): 131–39.

Trudgill, S. T., H. A. Viles, R. U. Cooke, R. J. Inkpen, A. L. Heathwaite, and J. Houston. 1992. Erratum: Trends in stone weathering and atmospheric pollution at St Paul's Cathedral, London, 1980–1990 (*Atmospheric Environment* 25A [12; 1991]: 2851–53). *Atmospheric Environment—Part A General Topics* 26 (9): 1801.

Trudgill, S. T., H. A. Viles, R. Inkpen, C. Moses, W. Gosling, T. Yates, P. Collier, D. I. Smith, and R. U. Cooke. 2001. Twenty-year weathering remeasurements at St Paul's Cathedral, London. *Earth Surface Processes and Landforms* 26 (10): 1129–42.

TU Delft (Delft University of Technology). 2009. EU Project Desalination: Assessment of desalination mortars and poultices for historic masonry. www.citg.tudelft.nl/live/pagina.jsp?id=267cbaf8-92c8-4204-92c0-97c32fff7eb5&lang=en.

Turpin, S. A., R. P. Watson, S. Dennett, and H. Muessig. 1979. Stereophotogrammetric documentation of exposed archaeological features. *Journal of Field Archaeology* 6 (3): 329–37.

Uchida, E., Y. Ogawa, N. Maeda, and T. Nakagawa. 2000. Deterioration of stone materials in the Angkor monuments, Cambodia. *Engineering Geology* 55 (1–2): 101–12.

UNI (Ente nazionale italiano di unificazione/Italian Organization for Standardization). 2006. UNI 11182 *Beni culturali: Materiali lapidei naturali ed artificiali; Descrizione della forma di alterazione; Termini e definizioni = Cultural heritage: Natural and artificial stone; Description of the alteration; Terminology and definition.* Milan and Rome: UNI (Ente nazionale italiano di unificazione).

United States Environmental Protection Agency (US EPA). 1997. *Air Quality Criteria for Particulate Matter.* Research Triangle Park, N.C. US Environmental Protection Agency, Office of Research and Development, National Center for Environmental Assessment.

Urzï, C., and F. De Leo. 2001. Sampling with adhesive tape strips: An easy and rapid method to monitor microbial colonization on monument surfaces. *Journal of Microbiological Methods* 44 (1): 1–11.

Valdeón, L., M. S. King, and M. H. De Freitas. 1992. Ultrasonic methods for quantifying the degradation of building stones. In *Proceedings of the 7th International Congress on Deterioration and Conservation of Stone, Lisbon, 15–18 June 1992,* ed. J. Delgado Rodrigues, F. Henriques, and F. Telmo Jeremias, 697–704. Lisbon: Laboratório Nacional de Engenharia Civil.

Vallet, J.-M., P. Bromblet, V. Vergès-Belmin, I. Pallot-Fossard, and F. Henry. 2000. *La protection des pierres: Guide sur les hydrofuges de surface.* Les cahiers techniques du cercle des partenaires du patrimoine 3. Champs-sur-Marne: Cercle des partenaires du patrimoine.

Vallet, J.-M., C. Gosselin, P. Bromblet, O. Rolland, V. Vergès-Belmin, and W. Kloppmann. 2006. Origin of salts in stone monument degradation using sulphur and oxygen isotopes: First results of the Bourges cathedral (France). *Journal of Geochemical Exploration* 88 (1–3): 358–62.

Van Balen, K. 1996. Expert system for evaluation of deterioration of ancient brick masonry structures. *Science of the Total Environment* 189–190: 247–54.

Van Balen, K. 1999. *Expert System for the Evaluation of the Deterioration of Ancient Brick Structures: Scientific Background of the Damage Atlas and the Masonry Damage Diagnostic System.* Research Report (European Commission, Directorate-General XII, Science, Research, and Development) 8. Luxembourg: Office for Official Publications of the European Communities, and Lanham, MD: Bernan Associates (distributor).

Van Balen, K. E. P., S. Ercan, and T. C. Patricio. 1999. Compatibility and retreatability versus reversibility: A case study at the late Hellenistic Nymphaeum of Sagalassos (Turkey). In *The Use of and Need for Preservation Standards in Architectural Conservation,* ed. L. B. Sickles-Taves, 105–18. ASTM Special Technical Publication 1355. W. Conshohocken, PA: ASTM.

Vandevoorde, D., M. Pamplona, O. Schalm, Y. Vanhellemont, V. Cnudde, and E. Verhaeven. 2009. Contact sponge method: Performance of a promising tool for measuring the initial water absorption. *Journal of Cultural Heritage* 10 (1): 41–47.

Van Hees, R. P. J., ed. 1998. *Evaluation of the Performance of Surface Treatments for the Conservation of Historic Brick Masonry.* Protection and Conservation of the European Cultural Heritage Research Report 7. Luxembourg: European Communities.

Van Hees, R. P. J., and J. A. G. Koek. 1995. Treatment of rising damp: A laboratory evaluation method. In *Methods of Evaluating Products for the Conservation of Porous Building Materials in Monuments: International Colloquium, Rome, 19–21 June 1995; Preprints,* 403–18. Rome: ICCROM.

Van Hees, R., S. Naldini, and B. Lubelli. 2009. The development of MDDS-COMPASS. Compatibility of plasters with salt loaded substrates. *Construction and Building Materials* 23 (5): 1719–30.

Van Hees, R. P. J., S. Naldini, and M. Sanders. 2006. An expert system for analysis of damage to plasters due to salt and moisture. In *Sais solúveis em argamassas de edifícios antigos: Danos, processos e solucões, Lisboa, LNEC, 14 e 15 de Fevereiro de 2005,* 16.1–16.11. Cursos e seminários 32. Lisbon: Laboratório Nacional de Engenharia Civil.

Vazquez-Calvo, C., M. Alvarez de Buergo, and R. Fort. 2007. Overview of recent knowledge of patinas on stone monuments: The Spanish experience. In *Building Stone Decay: From Diagnosis to Conservation,* ed. R. Přikryl and B. J. Smith, 295–307. Geological Society Special Publication 271. London: Geological Society of London.

Venkataraman, B. 2008. Microbes eating away at pieces of history. *New York Times,* June 24. http://www.nytimes.com/2008/06/24/health/24iht-24micr.13933560.html.

Verdel, T., and C. Chambon. 1994. Environmental degradation study of the Sphinx (Egypt) by the use of system dynamics: A new approach at the disposal of specialists. In *Conservation of Monuments in the Mediterranean Basin: Stone Monuments, Methodologies for the Analysis of Weathering and Conservation; Proceedings of the 3rd International Symposium, Venice, 22–25 June 1994 = La conservazione dei monumenti nel bacino del Mediterraneo: Materiali lapidei e monumenti, metodologie per l'analisi del degrado e la conservazione; Atti del 3° simposio internazionale, Venezia, 22–25 giugno 1994,* ed. V. Fassina, H. Ott, and F. Zezza, 383–92. Venice: Soprintendenza ai beni artistici e storici di Venezia.

Vergès-Belmin, V. 1992. Répartition des sels et cartographie des altérations sur la façade de l'église Notre-Dame-La-Grande à Poitiers, France = Salts distribution and mapping of stone surface weathering on the facade of Notre-Dame-La-Grande in Poitiers, France. In *Proceedings of the 7th International Congress on Deterioration and Conservation of Stone, Lisbon, 15–18 June 1992,* ed. J. Delgado Rodrigues, F. Henriques, and F. Telmo Jeremias, 927–36. Lisbon: Laboratório Nacional de Engenharia Civil.

Vergès-Belmin, V. 1994. Pseudomorphism of gypsum after calcite: A new textural feature accounting for the marble sulphation mechanism. *Atmospheric Environment – Part A* 28 (2): 295–304.

Vergès-Belmin, V. 1996a. Nettoyage des pierres des monuments français par laser: comparaison avec d'autres méthodes, microsablage et compresse chimique. In *Conservation et restauration des biens culturels: Pierre, pollution atmosphérique, peinture murale, études scientifiques et cas pratiques; Actes du Congrès LCP 1995, Montreux 24–29 septembre 1995 = Preservation and Restoration of Cultural Heritage: Stone Materials, Air Pollution, Murals, Scientific Research Work and Case Studies; Proceedings of the 1995 LPC Congress, Montreux 24–29 September 1995,* ed. R. Pancella, 481–90. Lausanne: Laboratoire de conservation de la pierre, Département des matériaux, EPFL (Ecole polytechnique fédérale de Lausanne).

Vergès-Belmin, V. 1996b. Towards a definition of common evaluation criteria for the cleaning of porous building materials: A review. *Science and Technology for Cultural Heritage* 5 (1): 69–83.

Vergès-Belmin, V., ed. 2008. *Illustrated Glossary on Stone Deterioration Patterns = Glossaire illustré sur les formes d'altération de la pierre*. English-French ed. Monuments & Sites 15. Paris: ICOMOS (International Council on Monuments and Sites) and ISCS (International Scientific Committee for Stone). http://international.icomos.org/publications/monuments_and_sites/15/pdf/Monuments_and_Sites_15_ISCS_Glossary_Stone.pdf.

Vergès-Belmin, V., and P. Bromblet. 2000. Le nettoyage de la pierre. *Monumental: Revue scientifique et technique de la Sous-direction des monuments historiques* (2000): 220–73.

Vergès-Belmin, V., and C. Dignard. 2003. Laser yellowing: Myth or reality? *Journal of Cultural Heritage* 4 (Suppl. 1: Lasers in the Conservation of Artworks—LACONA IV): 238–44.

Vergès-Belmin, V., and M. Laboure. 2007. Poultices as a way to eliminate the yellowing effect linked to limestone laser cleaning. In *Lasers in the Conservation of Artworks: LACONA VI, Vienna, Austria, Sept. 21–25, 2005; Proceedings,* ed. J. Nimmrichter, W. Kautek, and M. Schreiner, 115–24. Springer Proceedings in Physics 116. Berlin and New York: Springer.

Vergès-Belmin, V., and H. Siedel. 2005. Desalination of masonries and monumental sculptures by poulticing: A review. *Restoration of Buildings and Monuments: An International Journal = Bauinstandsetzen und Baudenkmalpflege: Eine internationale Zeitschrift* 11: 391–408.

Vergès-Belmin, V., C. Pichot, and G. Orial. 1994. Use of petrography for the comparison of laser-beam and microsandblasting cleaning techniques. In *EUROCARE-EUROMARBLE EU 496: Proceedings of the Fourth Workshop, Arles, France, November 3–6, 1993,* ed. R. Snethlage and V. Vergès-Belmin. Forschungsbericht (Bayerisches Landesamt für Denkmalpflege) 13. Munich: Bayerisches Landesamt für Denkmalpflege.

Vergès-Belmin, V., O. Rolland, and L. Leroux. 2008. Can we be confident in colour measurements performed outdoors? In *Proceedings of the 11th International Congress on Deterioration and Conservation of Stone, 15–20 September 2008, Toruń, Poland,* ed. J. W. Lukaszewicz and P. Niemcewicz, 539–46. Toruń, Poland: Nicolaus Copernicus University.

Vergès-Belmin, V., G. Wiedemann, L. Weber, M. Cooper, D. Crump, and R. Gouerne. 2003. A review of health hazards linked to the use of lasers for stone cleaning. *Journal of Cultural Heritage* 4 (Suppl. 1: Lasers in the Conservation of Artworks—LACONA IV): 33–37.

Viles, H. A. 1997. Urban air pollution and the deterioration of buildings and monuments. In *The Global Environment: Science, Technology and Management,* ed. D. Brune, D. V. Chapman, M. D. Gwynne, and J. M. Pacyna, 599–609. [Oslo]: Scandinavian Science Publisher and Weinheim, VCH.

Viles, H. 2003. Book review: Interdisciplinary Studies in Ancient Stone: ASMOSIA 5 by John J. Herrmann Jr., Norman Herz, and Richard Newman, editors. *Studies in Conservation* 48 (2): 142–43.

Viles, H. 2005. Can stone decay be chaotic? In *Stone Decay in the Architectural Environment,* ed. A. V. Turkington, 11–16. GSA Special Paper 390. Boulder, CO: Geological Society of America.

Viles, H. A., D. Camuffo, S. Fitz, B. Fitzner, O. Lindqvist, R. A. Livingston, P. V. Maravellaki, C. Sabbioni, and T. Warscheid. 1997. Group report: What is the state of our knowledge of the mechanisms of deterioration and how good are our estimates of rates of deterioration? In *Saving Our Architectural Heritage: The Conservation of Historic Stone Structures; Report of the Dahlem Workshop on Saving Our Architectural Heritage, The Conservation of Historic Stone Structures, Berlin, March 3–8, 1996*, ed. N. S. Baer and R. Snethlage, 95–112. Dahlem Workshop Reports. Chichester and New York: John Wiley & Sons.

Villegas, R., and J. F. Vale. 1992. Evaluation of the behaviour of water repellent treatments for stone. In *Proceedings of the 7th International Congress on Deterioration and Conservation of Stone, Lisbon, 15–18 June 1992,* ed. J. Delgado Rodrigues, F. Henriques, and F. Telmo Jeremias, vol. 3, 1253–62. Lisbon: Laboratório Nacional de Engenharia Civil.

Villegas, R., J. F. Vale, and Miguel A. Bello. 1994. Evaluation of consolidants by means of ultrasonic and surface hardness measurements. In *Conservation of Monuments in the Mediterranean Basin: Stone Monuments, Methodologies for the Analysis of Weathering and Conservation; Proceedings of the 3rd International Symposium, Venice, 22–25 June 1994 = La conservazione dei monumenti nel bacino del Mediterraneo: Materiali lapidei e monumenti, metodologie per l'analisi del degrado e la conservazione; Atti del 3° simposio internazionale, Venezia, 22–25 giugno 1994,* ed. V. Fassina, H. Ott, and F. Zezza, 919–23. Venice: Soprintendenza ai beni artistici e storici di Venezia.

Vlassenbroeck, J., V. Cnudde, B. Masschaele, M. Dierick, L. Van Hoorebeke, and P. Jacobs. 2007. A comparative and critical study of X-ray CT and neutron CT as non-destructive material evaluation techniques. In *Building Stone Decay: From Diagnosis to Conservation,* ed. R. Přikryl and B. J. Smith, 277–85. Geological Society Special Publication 271. London: Geological Society of London.

Vleugels, G., R. Dewolfs, and R. Van Grieken. 1993. On the memory effect of limestone for air pollution. *Atmospheric Environment—Part A General Topics* 27A (12): 1931–34.

Wainwright, I. N. M., H. Sears, and S. Michalski. 1997. Design of a rock art protective structure at Petroglyphs Provincial Park, Ontario, Canada. *Journal of the Canadian Association for Conservation = Journal de l'Association canadienne pour la conservation et la restauration* 22: 53–76.

Wakefield, R. D., and M. S. Jones. 1998. An introduction to stone colonizing microorganisms and biodeterioration of building stone. *Quarterly Journal of Engineering Geology* 31 (4, pt. 4): 301–13.

Walderhaug, O. 1998. Chemical weathering at rock art sites in western Norway: Which mechanisms are active and how can they be retarded? *Journal of Archaeological Science* 25 (8): 789–800.

Walderhaug, O., and E. M. Walderhaug. 1998. Weathering of Norwegian rock art: A critical review. *Norwegian Archaeological Review* 31: 119–39.

Walderhaug Saetersdal, E. M. 2000. Ethics, politics and practices in rock art conservation. *Public Archaeology* 1 (3): 163–80.

Walker, J. J., and N. R. Pace. 2007. Endolithic microbial ecosystems. *Annual Review of Microbiology* 61: 331–47. http://arjournals.annualreviews.org/doi/abs/10.1146/annurev.micro.61.080706.093302.

Wangler, T., and G. W. Scherer. 2008. Clay swelling mechanism in clay-bearing sandstones. *Environmental Geology* 56 (3–4): 529–34.

Ward, G., and L. A. Ward, eds. 1996. *Management of Rock Imagery*. AURA Occasional Publication 9. Melbourne: Australian Rock Art Research Association.

Warke, P. A., and B. J. Smith. 1998. Effects of direct and indirect heating on the validity of rock weathering simulation studies and durability tests. *Geomorphology* 22 (3–4): 347–57.

Warrack, S. 2000. The documentation of the conservation of the sandstone reliefs at Angkor Wat in Cambodia. In *GraDoc (Graphic Documentation Systems in Mural Painting Conservation) Research Seminar, Rome 16–20 November 1999,* ed. W. Schmid, 166–68. Rome: ICCROM.

Warscheid, T. 2008. Heritage research and practice: Towards a better understanding? In *Heritage Microbiology and Science: Microbes, Monuments and Maritime Materials,* ed. E. May, M. Jones, and J. Mitchell, 11–26. Royal Society of Chemistry (Great Britain) Special Publication 315. Cambridge: Royal Society of Chemistry. http://www.rsc.org/ebooks/2008/BK9780854041411/BK9780854041411-00011.pdf.

Warscheid, T., and J. Braams. 2000. Biodeterioration of stone: A review. *International Biodeterioration and Biodegradation* 46 (4): 343–68.

Watchman, A. 1998. Composition and source of dust on Split Rock paintings, Australia. *Rock Art Research: The Journal of the Australian Rock Art Research Association (AURA)* 15 (1): 36–40.

Watchman, A. 2005. Conservation of Australian rock art. *Coalition: CSIC Thematic Network on Cultural Heritage Electronic Newsletter* 10 (July): 14–18. http://www2.cica.es/ListasD/coalition/attachments/20050801/e0dba034/NL10-0001.obj.

Webb, A. H., R. J. Bawden, A. K. Busby, and J. N. Hopkins. 1992. Studies on the effects of air pollution on limestone degradation in Great Britain. *Atmospheric Environment – Part B Urban Atmosphere* 26 (2): 165–81.

Webster, A., and E. May. 2006. Bioremediation of weathered-building stone surfaces. *Trends in Biotechnology* 24 (6): 255–60.

Webster, A. M., D. Vicente, and E. May. 2004. Bacteria and the bioremedition of stone: The potential for saving cultural heritage. In *Environmental Biotechnology - ESEB 2004: Proceedings of the European Symposium on Environmental Biotechnology, 25–28 April 2004, Oostende, Belgium,* ed. W. Verstraete, 793–96. Leiden, London, and New York: Balkema.

Weiss, T., P. N. J. Rasolofosaon, and S. Siegesmund. 2002. Ultrasonic wave velocities as a diagnostic tool for the quality assessment of marble. In *Natural Stone, Weathering Phenomena, Conservation Strategies and Case Studies,* ed. S. Siegesmund, T. Weiss, and A. Vollbrecht, 149–64. Geological Society Special Publication 205. London: Geological Society of London.

Weiss, T., S. Siegesmund, D. Kirchner, and J. Sippel. 2004. Insolation weathering and hygric dilatation: Two competitive factors in stone degradation. *Environmental Geology* 46 (3–4): 402–13.

Werner, M. 1991. Research on cleaning methods applied to historical stone monuments. In *Science, Technology, and European Cultural Heritage: Proceedings of the European Symposium, Bologna, Italy, 13–16 June 1989,* ed. N. S. Baer, C. Sabbioni, and A. I. Sors, 688–91. Oxford and Boston: Published for the Commission of the European Communities by Butterworth-Heinemann Publishers.

Wheeler, G. 2008. Alkoxysilanes and the consolidation of stone: Where we are now. In *Stone Consolidation in Cultural Heritage: Research and Practice; Proceedings of the International Symposium, Lisbon, 6–7 May 2008,* ed. J. Delgado Rodrigues and J. M. Mimoso, 41–52. Lisbon: LNEC (Laboratório Nacional de Engenharia Civil).

Wheeler, G., and E. S. Goins. 2005. *Alkoxysilanes and the Consolidation of Stone.* Research in Conservation. Los Angeles: Getty Conservation Institute.

Wheeler, G., J. Mendez-Vivar, and S. Fleming. 2003. The use of modified Zr-npropoxide in the consolidation of calcite: A preliminary study focused into the conservation of cultural heritage. *Journal of Sol-Gel Science and Technology* 26 (1–3): 1233–37.

Wheeler, G. S., E. Wolkow, and H. Gafney. 1992. Microstructures of B72 acrylic resin/MTMOS composites. In *Materials Issues in Art and Archaeology III Symposium, Held April 27–May 1, 1992, San Francisco,* ed. P. B. Vandiver, J. R. Druzik, G. S. Wheeler, and I. C. Freestone, 963–67. Materials Research Society Symposium Proceedings 267. Pittsburgh: Materials Research Society.

Wheeler, G. S., G. L. Shearer, S. Fleming, L. W. Kelts, A. Vega, and R. J. Koestler. 1991. Toward a better understanding of B72 acrylic resin/methyltrimethoxysilane stone consolidants. In *Materials Issues in Art and Archaeology II Symposium, Held April 17–21, 1990, San Francisco,* ed. P. B. Vandiver, J. R. Druzik, and G. S. Wheeler, 209–26. Materials Research Society Symposium Proceedings 185. Pittsburgh: Materials Research Society.

Whitley, D. S., ed. 2001. *Handbook of Rock Art Research.* Walnut Creek, CA: AltaMira Press.

Whitley, D. S. 2005. *Introduction to Rock Art Research*. Walnut Creek, CA: Left Coast Press.

Whitley, D. S. 2006. U.S. rock art in the twenty-first century: Problems and prospects. *Conservation: The GCI Newsletter* 21 (3). http://www.getty.edu/conservation/publications/newsletters/21_3/news_in_cons1.html.

Wilson, M. J. 2004. Weathering of rocks by lichens with special reference to stonework: A review. *Stone Deterioration in Polluted Urban Environments,* ed. D. J. Mitchell and D. E. Searle, 51–60. Land Reconstruction and Management Series 3. Enfield, NH: Science Publishers.

Winkler, E. M. 1975. *Stone, Properties, Durability in Man's Environment*. 2nd rev. ed. Applied Mineralogy 4. New York: Springer-Verlag.

Winkler, E. M. 1985. A durability index for stone. In *Ve Congrès international sur l'altération et la conservation de la Pierre, Lausanne, 25–27.9.1985: Actes = 5th International Congress on Deterioration and Conservation of Stone, Lausanne, 25–27.9.1985: Proceedings,* ed. G. Félix, 151–56. Lausanne: Presses polytechniques romandes.

Wissenschaftlich-Technische-Arbeitsgemeinschaft für Bauwerkserhaltung und Denkmalpflege e.V. 1999. *WTA Mauerwerkdiagnostik Merkblatt 4–5–99/D*. München: WTA-Publications.

Wittmann, F. H., Y.-Z. Xian, T.-J. Zhao, and S. Giessler-Blank. 2008. Moisture diffusion and siloxane distribution in integral water repellent concrete. In *Restoration of Buildings and Monuments: An International Journal = Bauinstandsetzen und Baudenkmalpflege: Eine internationale Zeitschrift* 14 (1): 15–26.

Woolf, C. J. 1992. Book Review of *Dahlem Workshop, Life Sciences Research Report,* Volume 49, 1991. In *Brain: A Journal of Neurology* 115: 641-42.

Woolfitt, C., and G. Abrey. 2000. Poultices: The true or plain poultice and the cleaning and desalination of historic masonry and sculpture. *The Building Conservation Directory*. http://www.buildingconservation.com/articles/poultices/poultice.htm.

Woolfitt, C., and N. Durnan. 2002. Lime method evaluation: A survey of sites where lime-based conservation techniques have been employed to treat decaying limestone. In *Stone: Stone Building Materials, Construction and Associated Component Systems; Their Decay and Treatment,* ed. J. Fidler, 29–44. English Heritage Research Transactions 2. London: James & James.

World Health Organization (WHO). 2008. Sulfur dioxide (SO_2): Guideline values. In *Air Quality and Health,* Fact Sheet. no. 313. Copenhagen: World Health Organization. http://www.who.int/mediacentre/factsheets/fs313/en/index.html.

Worth, D. M. 2007. Book review: Cleaning Techniques in Conservation Practice. *APT Bulletin* 38 (2–3): 68.

Wu, L., A. D. Jacobson, and M. Hausner. 2008. Characterization of elemental release during microbe-granite interactions at T = 28°C. *Geochimica et Cosmochimica Acta* 72 (4): 1076–95.

Wu, L., A. D. Jacobson, H.-C. Chen, and M. Hausner. 2007. Characterization of elemental release during microbe-basalt interactions at T = 28°C. *Geochimica et Cosmochimica Acta* 71 (9): 2224–39.

Wüst, R. A. J., and C. Schlüchter. 2000. The origin of soluble salts in rocks of the Thebes mountains, Egypt: The damage potential to ancient Egyptian wall art. *Journal of Archaeological Science* 27 (12): 1161–72.

Xiangmin Zhang, and C. J. Spiers. 2005. Effects of phosphate ions on intergranular pressure solution in calcite: An experimental study. *Geochimica et Cosmochimica Acta* 69 (24): 5681–91.

Xingang Liu, Yafang Cheng, Yuanhang Zhang, Jinsang Jung, Nobuo Sugimoto, Shih-Yu Chang, Young J. Kim, Shaojia Fan, and Limin Zeng. 2008. Influences of relative humidity and particle chemical composition on aerosol scattering properties during the 2006 PRD campaign. *Atmospheric Environment* 42 (7): 1525–36.

Yang, M., G. W. Scherer, and G. S. Wheeler. 1998. Compatible consolidants. In *Compatible Materials for the Protection of European Cultural Heritage,* ed. G. Biscontin, A. Moropoulou, M. Erdik, and J. Delgado Rodrigues, vol. 2, 201–8. PACT 56. Athens: Technical Chamber of Commerce.

Yerrapragada, S. S., S. S. Tambe, and K. L. Gauri. 1993. Fractals, pore potential, and Sphinx limestone durability. In *Rock for Erosion Control,* ed. C. H. McElroy and D. A. Lienhart, 38–45. ASTM Special Technical Publication 1177. Philadelphia: ASTM.

Yilmaz, I. 2008. Discussion on the paper entitled "Mechanical and physical properties of the Kandira stone, Kandira, Turkey" by Arman et al., Bull Eng Geol Environ. *Bulletin of Engineering Geology and the Environment* 67 (2): 289–90. DOI: 10.1007/s10064-006-0082-x.

Young, A. R. M. 1987. Salt as an agent in the development of cavernous weathering. *Geology* 15 (10): 962–66.

Young, M. E. 1993. Stone cleaning of granite buildings and monuments: The use of colour measurement in stone cleaning trials. In *Conservation Science in the U.K.: Preprints of the Meeting Held in Glasgow, May 1993,* ed. N. H. Tennent, 20–28. London: James & James Science Publishers.

Young, P. 1996. Pollution-fueled "biodeterioration" threatens historic stone. *Environmental Science and Technology* 30 (5): 206A–8A.

Young, D., and D. Ellsmore. 2008. *Salt Attack and Rising Damp: A Guide to Salt Damp in Historic and Older Buildings.* 2nd ed. Sydney: Heritage Council of NSW, Heritage Victoria, South Australian Dept. for Environment and Heritage, and Adelaide City Council. http://www.heritage.nsw.gov.au/docs/HVC014_Salt_Damp_tech_guide_FA_web.pdf.

Young, M., and D. Urquhart. 1992. Abrasive cleaning of sandstone buildings and monuments: An experimental investigation. In *Stone Cleaning and the Nature, Soiling and Decay Mechanisms of Stone: Proceedings of the International Conference Held in Edinburgh, UK, 14–16 April 1992,* ed. R. G. M. Webster, 128–40. London: Donhead Publishing.

Young, G. S., and I. N. M. Wainwright. 1995. The control of algal biodeterioration of a marble petroglyph site. *Studies in Conservation* 40 (2): 82–92.

Young, M. E., D. C. M. Urquhart, and R. A. Laing. 2003. Maintenance and repair issues for stone cleaned sandstone and granite building façades. *Building and Environment* 38 (9–10): 1125–31.

Young, M., J. Ball, R. A. Laing, and D. C. M. Urquhart. 2003. *Maintenance and Repair of Cleaned Stone Buildings.* Historic Scotland Technical Advice Notes 25. Edinburgh: Historic Scotland.

Yu, S., and C. T. Oguchi. 2009. Complex relationships between salt type and rock properties in a durability experiment of multiple salt-rock treatments. *Earth Surface Processes and Landforms* 34 (15): 2096–110.

Zádor, M. 1992. Consolidation of porous stones by weather resistant agents = Consolidation des pierres poreuses par des enduits stables. In *Proceedings of the 7th International Congress on Deterioration and Conservation of Stone, Lisbon, 15–18 June 1992,* ed. J. Delgado Rodrigues, F. Henriques, and F. Telmo Jeremias, 1289–98. Lisbon: Laboratório Nacional de Engenharia Civil.

Zafiropulos, V., C. Balas, A. Manousaki, Y. Marakis, P. Maravelaki-Kalaitzaki, K. Melesanaki, P. Pouli, T. Stratoudaki, S. Klein, J. Hildenhagen, K. Dickmann, B. S. Luk'Yanchuk, C. Mujat, and A. Dogariu. 2003. Yellowing effect and discoloration of pigments: Experimental and theoretical studies. *Journal of Cultural Heritage* 4 (Suppl. 1: Lasers in the Conservation of Artworks—LACONA IV): 245–56.

Zamarreño, D. V., R. Inkpen, and E. May. 2009. Carbonate crystals precipitated by freshwater bacteria and their use as a limestone consolidant. *Applied and Environmental Microbiology* 75 (18): 5981–90.

Zammit, G., F. De Leo, P. Albertano, and C. Urzì. 2008. A preliminary study of microbial communities colonizing ochre-decorated chambers at the Hal Saflieni Hypogeum at Paola, Malta. In *Proceedings of the 11th International Congress on Deterioration and Conservation of Stone, 15–20 September 2008, Toruń, Poland,* ed. J. W. Lukaszewicz and P. Niemcewicz, 555–62. Toruń, Poland: Nicolaus Copernicus University.

Zehnder, K., and O. Schoch. 2009. Efflorescence of mirabilite, epsomite and gypsum traced by automated monitoring on-site. *Journal of Cultural Heritage* 10 (3): 319–30.

Zezza, F. 1990. Computerized analysis of stone decay in monuments. In *Advanced Workshop: "Analytical Methodologies for the Investigation of Damaged Stones," Pavia, 14–21 September 1990*, ed. F. Veniale and U. Zezza, 163–84. Pavia, Italy: Dipartimento di scienze della terra, Sezione di mineralogia, petrografia, geochimica, Università degli studi di Pavia.

Zezza, F. 1994. Stone decay diagnosis and control of treatments by computerized analytical techniques. In *Conservation of Monuments in the Mediterranean Basin: Stone Monuments, Methodologies for the Analysis of Weathering and Conservation; Proceedings of the 3rd International Symposium, Venice, 22–25 June 1994 = La conservazione dei monumenti nel bacino del Mediterraneo: Materiali lapidei e monumenti, metodologie per l'analisi del degrado e la conservazione; Atti del 3° simposio internazionale, Venezia, 22–25 giugno 1994*, ed. V. Fassina, H. Ott, and F. Zezza, 77–81 and plates. Venice: Soprintendenza ai beni artistici e storici di Venezia.

Zezza, F. 2002. Non-destructive technique for the assessment of the deterioration processes of prehistoric rock art in karstic caves: The paleolithic paintings of Altamira (Spain). In *Protection and Conservation of the Cultural Heritage of the Mediterranean Cities*, ed. E. Galán Huertos and F. Zezza, 377–88. Lisse, Netherlands, and Exton, PA: Balkema.

Zhao Haiying, Li Zuixiong, Han Wenfeng, Sun Manli, and Wang Xudong. 2007. Study on the main disease of the Great Wall and its conservation in Gansu province. *Wen wu bao hu yu kao gu ke xue = Sciences of Conservation & Archaeology* 19 (1): 28–32.

Ziegenbalg, G. 2008. Colloidal calcium hydroxide: A new material for consolidation and conservation of carbonatic stones. In *Proceedings of the 11th International Congress on Deterioration and Conservation of Stone, 15–20 September 2008, Toruń, Poland*, ed. J. W. Lukaszewicz and P. Niemcewicz, 1109–15. Toruń, Poland: Nicolaus Copernicus University.

Zielecka, M., E. Bujnowska, and K. Bajdor. 2007. Siloxane-containing polymer matrices as coating materials. *Journal of Coatings Technology Research* 4 (3): 275–81.

Appendix: Resources for Stone Conservation

A wide range of publications was scrutinized during the writing of this book, and use was also made of databases such as AATA, BCIN, Scopus, ISI Web of Knowledge, and Science Direct; also, the Getty and UCLA libraries as well as many colleagues and other resources. Stone conservation is a multidisciplinary subject and thus relevant publications are to be found in many research areas. It is hoped that the following books and articles may provide the reader with a useful introduction to issues of research and application in the field of stone conservation.

This list is, by definition, incomplete and does not purport to be definitive but merely aspires to be useful. The emphasis of this listing, originally created to aid the students of the 2009 Venice Stone Course, is on stone, as mortars and grouts are covered elsewhere.

"Classic" Stone Conservation Texts

Ashurst, J., and N. Ashurst. 1988. *Practical Building Conservation.* Vol. 1, *Stone Masonry*. English Heritage Technical Handbook. New York: Halsted Press.

Ashurst, J., and F. G. Dimes, eds. 1998. *Conservation of Building and Decorative Stone*. Butterworth-Heinemann Series in Conservation and Museology. Oxford and Woburn, MA: Butterworth-Heinemann.

Lazzarini, L., and M. Laurenzi Tabasso. 1986. *Il restauro della pietra*. Padua: CEDAM.

Lazzarini, L., and M. Laurenzi Tabasso. 1989. *La restauration de la pierre*. Maurecourt: ERG.

Philippon, J., D. Jeannette, and R.-A. Lefèvre, eds. 1992. *La conservation de la pierre monumentale en France*. Conservation du patrimoine 2. Paris: Presses du CNRS, Ministère de la culture et de la communication.

Schaffer, R. J. 1932. *The Weathering of Natural Building Stones*. Department of Scientific and Industrial Research, Building Research Board, Special Report 18. London: H.M. Stationery Office [printed by Harrison and Sons].

Snethlage, R. 2008. *Leitfaden Steinkonservierung: Planung von Untersuchungen und Massnahmen zur Erhaltung von Denkmälern aus Naturstein*. 3rd ed. Stuttgart: IRB Verlag.

Stone Conservation Manuals and Glossaries

Borrelli, E., and A. Urland. 1999. *ARC Laboratory Handbook*. Rome: ICCROM. http://www.iccrom.org/pdf/ICCROM_14_ARCLabHandbook00_en.pdf.

Stone Federation of Great Britain. 1991. *Natural Stone Glossary*. London: Linked Advertising & Marketing.

Svahn, H. 2006. *Non-Destructive Field Tests in Stone Conservation: Field and Laboratory Tests: Final Report for the Research and Development Project*.

Rapport från Riksantikvarieämbetet 2006:4. Stockholm: Riksantikvarieämbetet. http://www.raa.se/publicerat/9172094354.pdf.

UNI (Ente nazionale italiano di unificazione/Italian Organization for Standardization). 2006. UNI 11182 Beni culturali: Materiali lapidei naturali ed artificiali; Descrizione della forma di alterazione; Termini e definizioni = Cultural heritage: Natural and artificial stone; Description of the alteration; Terminology and definition. Milan and Rome: UNI (Ente nazionale italiano di unificazione).

Vergès-Belmin, V., ed. 2008. *Illustrated Glossary on Stone Deterioration Patterns = Glossaire illustré sur les formes d'altération de la pierre*. English-French ed. Monuments & Sites 15. Paris: ICOMOS (International Council on Monuments and Sites) and ISCS (International Scientific Committee for Stone). http://international.icomos.org/publications/monuments_and_sites/15/pdf/Monuments_and_Sites_15_ISCS_Glossary_Stone.pdf.

Stone Conservation Overviews

Amoroso, G. G., and V. Fassina. 1983. *Stone Decay and Conservation: Atmospheric Pollution, Cleaning, Consolidation, and Protection*. Materials Science Monographs 11. Amsterdam and New York: Elsevier.

Baer, N. S., and R. Snethlage, eds. 1997. *Saving Our Architectural Heritage: The Conservation of Historic Stone Structures; Report of the Dahlem Workshop on Saving Our Architectural Heritage, The Conservation of Historic Stone Structures, Berlin, March 3–8, 1996*. Dahlem Workshop Reports. Chichester and New York: John Wiley & Sons.

Doehne, E., and J. L. Drever. 1994. Appendix I: Report on the impromptu discussion session on stone conservation. In *Durability and Change: The Science, Responsibility, and Cost of Sustaining Cultural Heritage; December 6–11, 1992*, ed. W. E. Krumbein, P. Brimblecombe, D. E. Cosgrove, and S. Staniforth, 287–89. Dahlem Workshop Reports: Environmental Sciences Research Report ES 15. Chichester and New York: John Wiley.

Domaslowski, W., ed. 2003. *Preventive Conservation of Stone Historical Objects*. Toruń, Poland: Nicolaus Copernicus University.

Fassina, V., ed. 2000. *Proceedings of the 9th International Congress on Deterioration and Conservation of Stone, Venice, June 19–24, 2000*. Amsterdam and New York: Elsevier.

Gauri, K. L., and J. K. Bandyopadhyay. 1999. *Carbonate Stone: Chemical Behavior, Durability, and Conservation*. New York: Wiley.

Henry, A. 2006. *Stone Conservation: Principles and Practice*. Shaftesbury, UK: Donhead Publishing.

Kwiatkowski, D., and R. Löfvendahl, eds. 2004. *Proceedings of the 10th International Congress on Deterioration and Conservation of Stone, Stockholm, June 27–July 2, 2004*. Stockholm: ICOMOS Sweden.

Lukaszewicz, J. W., and P. Niemcewicz, eds. 2008. *Proceedings of the 11th International Congress on Deterioration and Conservation of Stone, 15–20 September 2008, Toruń, Poland*. Toruń, Poland: Nicolaus Copernicus University.

Price, C. A. 1996. *Stone Conservation: An Overview of Current Research*. Research in Conservation. Santa Monica, CA: Getty Conservation Institute. http://www.getty.edu/conservation/publications/pdf_publications/stoneconservation.pdf.

Přikryl, R., and B. J. Smith, eds. 2007. *Building Stone Decay: From Diagnosis to Conservation*. Geological Society Special Publication 271. London: Geological Society of London.

Riederer, Josef, ed. 1996. *Proceedings of the 8th International Congress on Deterioration and Conservation of Stone, Berlin, 30 Sept.–4 Oct. 1996*. Berlin: Germany: Möller Druck und Verlag.

Siegesmund, S., R. Snethlage, and J. Ruedrich. 2008. Monument futures: Climate change, air pollution, decay and conservation; The Wolf-Dieter Grimm-volume. *Environmental Geology* 56 (3–4): 451–53.

Siegesmund, S., T. Weiss, and A. Vollbrecht, eds. 2002. *Natural Stone, Weathering Phenomena, Conservation Strategies and Case Studies*. Geological Society Special Publication 205. London: Geological Society of London.

Smith, B. J., and A. V. Turkington, eds. 2004. *Stone Decay: Its Causes and Controls; Proceedings of Weathering 2000, an International Symposium Held in Belfast, 26–30 June 2000* [SWAPNET: Stone Weathering and Atmospheric Pollution NETwork Conference]. Shaftesbury, UK: Donhead Publishing.

Teutonico, J.-M., and J. Fidler. 1998. Time for change: An overview of building-materials research for conservation of historic structures. *APT Bulletin* 29 (3–4): 45–49.

Thiel, M.-J., ed. 1993. *Conservation of Stone and Other Materials: Proceedings of the International RILEM/UNESCO Congress "Conservation of Stone and Other Materials: Research – Industry – Media," UNESCO Headquarters, Paris, June 29–July 1, 1993*. RILEM Proceedings 21. London and New York: E & FN Spon.

Torraca, G. 1999. The scientist in conservation. *Conservation: The GCI Newsletter* 14 (3). http://www.getty.edu/conservation/publications/newsletters/14_3/feature1_3.html.

Introduction to Stone Conservation Treatments and Practice

Cameron, S. 1997. *Biological Growths on Sandstone Buildings: Control and Treatment*. Historic Scotland Technical Advice Notes 10. Edinburgh: Historic Scotland.

Clifton, J. R. 1980. *Stone Consolidating Materials: A Status Report*. NBS Technical Note 1118. Washington, DC: U.S. Dept. of Commerce, National Bureau of Standards; Springfield, VA: National Technical Information Service (distributor). http://cool.conservation-us.org/byauth/clifton/stone/.

Delgado Rodrigues, J., and J. M. Mimoso, eds. 2008. *Stone Consolidation in Cultural Heritage: Research and Practice; Proceedings of the International Symposium, Lisbon, 6–7 May 2008*. Lisbon: LNEC (Laboratório Nacional de Engenharia Civil).

Feilden, B. M. 2003. *Conservation of Historic Buildings*. 3rd ed. Oxford: Architectural Press.

Fidler, J., ed. 2002. *Stone: Stone Building Materials, Construction and Associated Component Systems; Their Decay and Treatment*. English Heritage Research Transactions 2. London: James & James.

Hansen, E., E. Doehne, J. Fidler, J. Larson, B. Martin, M. Matteini, C. Rodríguez-Navarro, E. Sebastián Pardo, C. Price, A. de Tagle, J.-M. Teutonico, and N. R. Weiss. 2003. A review of selected inorganic consolidants and protective treatments for porous calcareous materials. *Reviews in Conservation* 4: 13–25.

International Centre for the Study of the Preservation and the Restoration of Cultural Property. 1995. *Methods of Evaluating Products for the Conservation of Porous Building Materials in Monuments: International Colloquium, Rome, 19–21 June, 1995; Preprints*. Rome: ICCROM.

Laurenzi Tabasso, M. 2004. Products and methods for the conservation of stone: Problems and trends. In *Proceedings of the 10th International Congress on Deterioration and Conservation of Stone, Stockholm, June 27–July 2, 2004*, ed. D. Kwiatkowski and R. Löfvendahl, vol. 1, 269–82. Stockholm: ICOMOS Sweden.

Laurenzi Tabasso, M., and S. Simon. 2006. Testing methods and criteria for the selection/evaluation of products for the conservation of porous building materials. *Reviews in Conservation* 7: 67–82.

Martin, W. 1996. Stone consolidants: A review. In *A Future for the Past: A Joint Conference of English Heritage and the Cathedral Architects Association: 25–26 March 1994*, ed. J.-M. Teutonico, 30–50. London: James & James.

Scherer, G. W., R. Flatt, and G. Wheeler. 2001. Materials science research for the conservation of sculpture and monuments. *MRS Bulletin* 26 (1): 44–50.

Scherer, G. W., and G. S. Wheeler. 2009. Silicate consolidants for stone. *Key Engineering Materials* 391: 1–25.

Snethlage, R., and E. Wendler. 2000. Chemical conservation of stone. In *Ullmann's Encyclopedia of Industrial Chemistry*, ed. F. Ullmann. 6th ed. New York: Wiley.

U.S. Department of the Interior. 2004. *The Preservation of Historic Architecture: The U.S. Government's Official Guidelines for Preserving Historic Homes*. Guilford, CT: Lyons Press.

Weaver, M. E. 1997. Restoring stonework. In *Conserving Buildings: Guide to Techniques and Materials,* ed. M. E. Weaver and F. G. Matero, 58–98. Rev. ed. New York: John Wiley & Sons.

Wheeler, G., and E. S. Goins. 2005. *Alkoxysilanes and the Consolidation of Stone*. Research in Conservation. Los Angeles: Getty Conservation Institute.

Wood, C. 2003. *Stone Roofing: Conserving the Materials and Practice of Traditional Stone Slate Roofing in England*. English Heritage Research Transactions: Research and Case Studies in Architectural Conservation 9. London: James & James.

Introduction to Stone and Masonry

Adkins, J. 2004. *Moving Heavy Things*. Brooklyn, ME: Wooden Boat Books.

Ashurst, J., and F. G. Dimes. 1977. *Stone in Building: Its Use and Potential Today*. London: Architectural Press.

Doyle, P., T. Hughes, and I. Thomas, eds. 2008. *England's Heritage in Stone: Proceedings of a Conference, Tempest Anderson Hall, York, 15–17 March, 2005*. Folkestone, Kent: English Stone Forum.

Heyman, J. 1997. *The Stone Skeleton: Structural Engineering of Masonry Architecture*. Cambridge and New York: Cambridge University Press.

Hill, P. R., and J. C. E. David. 1995. *Practical Stone Masonry*. London: Donhead Publishing.

Jefferson, D., S. Hanna, B. Martin, and D. M. Jones. 2006. *Identifying and Sourcing Stone for Historic Building Repair: An Approach to Determining and Obtaining Compatible Replacement Stone*. Technical Advice Note (English Heritage). Swindon, UK: English Heritage. http://www.shropshiregeology.org.uk/building_stones/English_Heritage_Identifying_and_Sourcing_Stone.pdf.

McRaven, C. 1989. *Building with Stone*. Pownal, VT: Storey Communications.

Rockwell, P. 1993. *The Art of Stoneworking: A Reference Guide*. Cambridge and New York: Cambridge University Press.

Shadmon, A. 1996. *Stone: An Introduction*. 2nd ed. London: Intermediate Technology.

Warland, E. G. 2006 [1929]. *Modern Practical Masonry*. Shaftesbury, UK: Donhead Publishing.

Winkler, E. M. 1994. *Stone in Architecture: Properties, Durability*. 3rd ed. Berlin and New York: Springer-Verlag.

Introduction to the History of Stone Production

ASMOSIA (Association for the Study of Marbles and Other Stones in Antiquity). ASMOSIA Proceedings. http://www.asmosia.org/pub_proc.html.

Heldal, T., E. Bloxam, and P. Storemyr. 2007. Unravelling ancient stone quarry landscapes in the Eastern Mediterranean: Three Egyptian case studies. In *Broadening Horizons: Multidisciplinary Approaches to Landscape Study,* ed. B. Ooghe and G. Verhoeven, 90–112. Newcastle, UK: Cambridge Scholars Publishing.

McKee, H. J. 1976. Brick and stone: Handicraft to machine. In *Building Early America: Contributions Toward the History of a Great Industry,* ed. C. E. Peterson, 74–95. Radnor, PA: Chilton Book Co.

Wurch-Kozelj, M. 1988. Methods of transporting blocks in antiquity. In *Classical Marble: Geochemistry, Technology, Trade,* ed. N. Herz and M. Waelkens, 55–64. ASMOSIA 1. NATO ASI Series: Series E, Applied Sciences 153. Dordrecht, Netherlands and Boston: Kluwer Academic Publishers.

Introduction to Science and Conservation

Moncrieff, A., G. Weaver, and J. Ashley-Smith, eds. 1992. *Science for Conservators*. Vol. 1, *An Introduction to Materials*. Conservation Science Teaching Series.

London and New York: The Conservation Unit of the Museums & Galleries Commission, in conjunction with Routledge.

Moncrieff, A., G. Weaver, and J. Ashley-Smith, eds. 1992. *Science for Conservators.* Vol. 2, *Cleaning*. Conservation Science Teaching Series. London and New York: The Conservation Unit of the Museums & Galleries Commission, in conjunction with Routledge.

Newey, C., R. Boff, V. Daniels, M. Pascoe, N. Tennant, and J. Ashley-Smith, eds. 1992. *Science for Conservators.* Vol. 3, *Adhesives and Coatings*. Conservation Science Teaching Series. London and New York: The Conservation Unit of the Museums & Galleries Commission, in conjunction with Routledge.

Torraca, G. 2009. *Lectures on Materials Science for Architectural Conservation.* Los Angeles: Getty Conservation Institute. http://www.getty.edu/conservation/publications/pdf_publications/torraca.pdf.

Introduction to Decay Mechanisms: Overview

Grassegger, G. 1999. Decay mechanisms of natural building stones on monuments: A review of the latest theories. In *Werkstoffe und Werkstoffprüfung im Bauwesen: Festschrift zum 60. Geburtstag von H.-W. Reinhardt,* ed. C. U. Große, 54–81. Stuttgart: IWB.

Smith, B. J., M. Gómez-Heras, and S. McCabe. 2008. Understanding the decay of stone-built cultural heritage. *Progress in Physical Geography* 32 (4): 439–61.

Introduction to Decay Mechanisms: Biodeterioration

Building Research Establishment. 1992. *Control of Lichens, Moulds and Similar Growths*. BRE Digest 370. Garston, Watford, UK: Building Research Establishment [Building Research Station].

Caneva, G., M. P. Nugari, and O. Salvadori, eds. 2008. *Plant Biology for Cultural Heritage: Biodeterioration and Conservation*. Los Angeles: Getty Conservation Institute.

Gorbushina, A. A., E. Diakumaku, L. Mueller, and W. E. Krumbein. 2003. Biocide treatment of rock and mural paintings: Problems of application, molecular techniques of control and environmental hazards. In *Molecular Biology and Cultural Heritage: Proceedings of the International Congress on Molecular Biology and Cultural Heritage, 4–7 March 2003, Sevilla, Spain,* ed. C. Saiz-Jimenez, 61–72. Lisse, Netherlands, and Exton, PA: Balkema.

Koestler, R. J. 1991. *Biodeterioration of Cultural Property*. London and New York: Elsevier Applied Science.

Nugari, M. P., and O. Salvadori. 2003. Biocides and treatment of stone: Limitations and future prospects. In *Art, Biology, and Conservation: Biodeterioration of Works of Art,* ed. R. J. Koestler, 519–35. New York: Metropolitan Museum of Art.

Walker, J. J., and N. R. Pace. 2007. Endolithic microbial ecosystems. *Annual Review of Microbiology* 61: 331–47.

Warscheid, T. 2008. Heritage research and practice: Towards a better understanding? In *Heritage Microbiology and Science: Microbes, Monuments and Maritime Materials,* ed. E. May, M. Jones, and J. Mitchell, 11–26. Royal Society of Chemistry (Great Britain) Special Publication 315. Cambridge: Royal Society of Chemistry. http://www.rsc.org/ebooks/2008/BK9780854041411/BK9780854041411-00011.pdf.

Warscheid, T., and J. Braams. 2000. Biodeterioration of stone: A review. *International Biodeterioration and Biodegradation* 46 (4): 343–68.

Introduction to Decay Mechanisms: Air Pollution and Environment

Brimblecombe, P., ed. 2003. *The Effects of Air Pollution on the Built Environment*. Air Pollution Reviews 2. London: Imperial College Press; Singapore and River Edge, NJ: World Scientific.

Charola, A. E., and R. Ware. 2002. Acid deposition and the deterioration of stone: A brief review of a broad topic. In *Natural Stone, Weathering Phenomena, Conservation Strategies and Case Studies,* ed. S. Siegesmund, T. Weiss, and A. Vollbrecht, 393–406. Geological Society Special Publication 205. London: Geological Society of London.

Cooke, R. U., and G. B. Gibbs. 1993. *Crumbling Heritage? Studies of Stone Weathering in Polluted Atmospheres: A Report of Research on Atmospheric Pollution and Stone Decay for the Joint Working Party Between the Cathedrals Fabric Commission for England and the Joint Environmental Programme of National Power Plc and Powergen Plc.* London: University College, Department of Geography.

Fassina, V. 1987. Environmental pollution in relation to stone decay. *Durability of Building Materials* 5: 317–58.

Livingston, R. A. 1997. Air pollution standards for architectural conservation. In *Saving Our Architectural Heritage: The Conservation of Historic Stone Structures; Report of the Dahlem Workshop on Saving Our Architectural Heritage, The Conservation of Historic Stone Structures, Berlin, March 3–8, 1996,* ed. N. S. Baer and R. Snethlage, 371–87. Dahlem Workshop Reports. Chichester and New York: John Wiley & Sons.

Meierding, T. C. 1993. Marble tombstone weathering and air pollution in North America. *Annals of the Association of American Geographers* 83 (4): 568–88.

Searle, D. E., and D. J. Mitchell. 2006. The effect of coal and diesel particulates on the weathering loss of Portland limestone in an urban environment. *Science of the Total Environment* 370 (1): 207–23.

Smith, B. J., and P. A. Warke, eds. 1996. *Processes of Urban Stone Decay: Proceedings of SWAPNET '95; Stone Weathering and Atmospheric Pollution Network Conference Held in Belfast, 19–20 May 1995.* London: Donhead Publishing.

Introduction to Decay Mechanisms: Salt and Frost Weathering

Arnold, A., and K. Zehnder. 1991. Monitoring wall paintings affected by soluble salts. In *The Conservation of Wall Paintings: Proceedings of a Symposium Organized by the Courtauld Institute of Art and the Getty Conservation Institute, London, July 13–16, 1987,* ed. S. Cather, 103–35. Marina del Rey, CA: Getty Conservation Institute. http://www.getty.edu/conservation/publications/pdf_publications/wall_paintings.pdf.

Borrelli, E. 1999. *Salts.* Vol. 3 of *ARC Laboratory Handbook,* ed. E. Borrelli and A. Urland. Rome: ICCROM. http://www.iccrom.org/pdf/ICCROM_14_ARCLabHandbook02_en.pdf.

Burt, T. P., R. J. Chorley, D. Brunsden, N. J. Cox, and A. S. Goudie, eds. 2008. *The History of the Study of Landforms: Or the Development of Geomorphology.* Vol. 4, *Quaternary and Recent Processes and Form (1890–1965) and the Mid-Century Revolution.* London: Geological Society Publishing House.

Charola, A. E., and J. Pühringer. 2005. Salts in the deterioration of porous materials: A call for the right questions. *Restoration of Buildings and Monuments: An International Journal = Bauinstandsetzen und Baudenkmalpflege: Eine internationale Zeitschrift* 11 (6): 433–42.

De Clercq, H. 2008. Performance of limestone contaminated with binary mixtures of sodium sulphate and treated with a water repellent. In *Hydrophobe V: Water Repellent Treatment of Building Materials; Proceedings of Hydrophobe V, Fifth International Conference on Water Repellent Treatment of Building Materials, Royal Institute for Cultural Heritage (KIK-IRPA), Brussels, Belgium, April 15 and 16, 2008,* ed. H. De Clercq and A. E. Charola, 107–15. Freiburg: Aedificatio Verlag.

Doehne, E. 2002. Salt weathering: A selective review. In *Natural Stone, Weathering Phenomena, Conservation Strategies and Case Studies,* ed. S. Siegesmund, T. Weiss, and A. Vollbrecht, 51–64. Geological Society Special Publication 205. London: Geological Society of London.

Matsuoka, N., and J. Murton. 2008. Frost weathering: Recent advances and future directions. *Permafrost and Periglacial Processes* 19 (2): 195–210.

Rodríguez-Navarro, C., and E. Doehne. 1999. Salt weathering: Influence of evaporation rate, supersaturation and crystallization pattern. *Earth Surface Processes and Landforms* 24 (2–3): 191–209.

Rossi-Manaresi, R., and A. Tucci. 1991. Pore structure and the disruptive or cementing effect of salt crystallization in various types of stone. *Studies in Conservation* 36 (1): 53–58.

Sawdy, A., A. Heritage, and L. Pel. 2008. A review of salt transport in porous media: Assessment methods and salt reduction treatments. In *Salt Weathering on Buildings and Stone Sculptures, 22–24 October 2008, The National Museum Copenhagen, Denmark* [Proceedings from the international conference], ed. J. S. Albertsen, 1–27. Lyngby: Technical University of Denmark, Department of Civil Engineering.

Scherer, G. W. 2004. Stress from crystallization of salt. *Cement and Concrete Research* 34 (9): 1613–24.

Scherer, G. W., and J. J. Valenza II. 2005. Mechanisms of frost damage. In *Materials Science of Concrete VII,* ed. F. Young and J. Skalny, vol. 7, 209–46. Westerville, OH: American Ceramic Society.

Young, D., and D. Ellsmore. 2008. *Salt Attack and Rising Damp: A Guide to Salt Damp in Historic and Older Buildings.* 2nd ed. Sydney: Heritage Council of NSW, Heritage Victoria, South Australian Dept. for Environment and Heritage, and Adelaide City Council. http://www.heritage.nsw.gov.au/docs/HVC014_Salt_Damp_tech_guide_FA_web.pdf.

Zehnder, K. 2007. Long-term monitoring of wall paintings affected by soluble salts. *Environmental Geology* 52 (2): 395–409.

Zehnder, K., and A. Arnold. 1989. Crystal growth in salt efflorescence. *Journal of Crystal Growth* 97 (2): 513–21.

Introduction to Stone Cleaning and Desalination

Allanbrook, T., and K. C. Normandin. 2007. The restoration of the Fifth Avenue facades of the Metropolitan Museum of Art. *APT Bulletin* 38 (4): 45–53.

Andrew, C. M., M. Young, and K. Tonge. 1994. *Stone Cleaning: A Guide for Practitioners.* Aberdeen: Robert Gordon University; Edinburgh: Historic Scotland.

Ashurst, J. 1998. The cleaning and treatment of limestone by the lime method. In *Conservation of Building and Decorative Stone,* ed. J. Ashurst and F. G. Dimes, vol. 2, 169–76. Butterworth-Heinemann Series in Conservation and Museology. London and Boston: Butterworth-Heinemann.

Ashurst, N. 1994. *Cleaning Historic Buildings.* London: Donhead Publishing.

British Standards Institution (BSI). 2000. *BS 8221-2:2000 Code of Practice for Cleaning and Surface Repair of Buildings: Surface Repair of Natural Stones, Brick and Terracotta.* London: BSI.

Clifton, J. R., ed. 1986. *Cleaning Stone and Masonry: A Symposium Sponsored by ASTM Committee E-6 on Performance of Building Constructions, Louisville, KY, 18 April 1983.* ASTM Special Technical Publication 935. Philadelphia: ASTM.

Cooper, M., ed. 1998. *Laser Cleaning in Conservation: An Introduction.* Woburn, MA: Butterworth-Heinemann.

Normandin, K. C., D. Slaton, N. R. Weiss, and J. Pearce, ed. 2005. Cleaning Techniques in Conservation Practice. Special issue, *Journal of Architectural Conservation* 11 (3).

Rodríguez-Navarro, C., K. Elert, E. Sebastián, R. M. Esbert, C. M. Grossi, A. Rojo, F. J. Alonso, M. Montoto, and J. Ordaz. 2003. Laser cleaning of stone materials: An overview of current research. *Reviews in Conservation* (4): 65–82.

Sawdy, A., A. Heritage, and L. Pel. 2008. A review of salt transport in porous media: Assessment methods and salt reduction treatments. In *Salt Weathering on Buildings and Stone Sculptures, 22–24 October 2008, The National Museum Copenhagen, Denmark* [Proceedings from the international conference], ed. J. S. Albertsen, 1–27. Lyngby: Technical University of Denmark, Department of Civil Engineering.

Young, M., J. Ball, R. A. Laing, and D. C. M. Urquhart. 2003. *Maintenance and Repair of Cleaned Stone Buildings*. Historic Scotland Technical Advice Notes 25. Edinburgh: Historic Scotland.

Introduction to Stone Monitoring, Moisture, and Environment

Camuffo, D. 1998. *Microclimate for Cultural Heritage*. Developments in Atmospheric Science 23. Amsterdam and New York: Elsevier.

Doehne, E., and S. Pinchin. 2008. Time-lapse macro-imaging in the field: Monitoring rapid flaking of magnesian limestone. In *Proceedings of the 11th International Congress on Deterioration and Conservation of Stone, 15–20 September 2008, Toruń, Poland,* ed. J. W. Lukaszewicz and P. Niemcewicz, 365–72. Toruń, Poland: Nicolaus Copernicus University.

Hall, C., and W. D. Hoff. 2002. *Water Transport in Brick, Stone, and Concrete*. London and New York: Spon Press.

Massari, G. and I. Massari. 1993. *Damp Buildings, Old and New.* ICCROM, International Centre for the Study of the Preservation and Restoration of Cultural Property, Rome. Rome: ICCROM.

Pender, R. J. 2004. The behaviour of water in porous building materials and structures. *Reviews in Conservation* (5): 49–62.

Thornbush, M. J., and H. A. Viles. 2008. Photographic monitoring of soiling and decay of roadside walls in central Oxford, England. *Environmental Geology* 56 (3–4): 777–87.

Tiano, P., and C. Pardini, eds. 2008. *In Situ Monitoring of Monumental Surfaces: Proceedings of the International Workshop SMW08, 27–29 October 2008, Florence, Italy*. Florence: Edifir.

Introduction to Rock Art Conservation

Dorn, R. I., D. S. Whitley, N. V. Cerveny, S. J. Gordon, C. D. Allen, and E. Gutbrod. 2008. The Rock Art Stability Index: A new strategy for maximizing the sustainability of rock art. *Heritage Management* 1 (1): 37–70.

Hygen, A.-S. 2006. *Protection of Rock Art, The Rock Art Project 1996–2005: Final Report from the Directorate for Cultural Heritage*. Oslo: Directorate for Cultural Heritage Archive. http://www.riksantikvaren.no/filestore/rockart project-finalreport.pdf.

Levin, J., ed. 2006. Rock Art. Special issue, *Conservation: The GCI Newsletter* 21 (3). http://getty.edu/conservation/publications/newsletters/21_3/index.html.

MacLeod, I. 2000. Rock art conservation and management: The past, present and future options. *Reviews in Conservation* 1: 32–45.

Saiz-Jimenez, C., ed. 2006. Conservation of Rock Art. Special issue, *COALITION: CSIC Thematic Network on Cultural Heritage Electronic Newsletter* (10–12). http://www.rtphc.csic.es.

Whitley, D. S., ed. 2001. *Handbook of Rock Art Research*. Walnut Creek, CA: Alta Mira Press.

Other important resources include the proceedings from numerous regional and international rock art conferences such as those held by the IFRAO (International Federation of Rock Art Organizations) and the Australian Rock Art Research Association (AURA). Some otherwise difficult-to-find material can be found in the California Rock Art Studies Bibliographic Database at the University of California, Berkeley (Marymor, L. 2009. Rock Art Studies: A Bibliographic Database. Available at http://bancroft.berkeley.edu/collections/rockart.html), although it is not up to date as of this writing. For the present overview, over four hundred bibliographic sources related to rock art conservation were collected and reviewed.

Introduction to the Conservation of Ornamental Stones

Fellowes, D., and P. Hagan. 2003. Pyrite oxidation: The conservation of historic shipwrecks and geological and palaeontological specimens. *Reviews in Conservation* (4): 26–38.

Jiménez-González, I., C. Rodríguez-Navarro, and G. W. Scherer. 2008. Role of clay minerals in the physicomechanical deterioration of sandstone. *Journal of Geophysical Research F: Earth Surface* 113 (2): F02021.

Siegesmund, S., J. Ruedrich, and A. Koch. 2008. Marble bowing: Comparative studies of three different public building facades. *Environmental Geology* 56 (3–4): 473–94.

Zehnder, K. 2006. Greying of black polished limestone: A case study to clarify the phenomenon. *Zeitschrift für Kunsttechnologie und Konservierung* 20 (2): 361–67.

Some Related EC Projects and Results

Bourgès, A., and V. Vergès-Belmin. 2008b. Comparison and optimization of five desalination systems on the inner walls of Saint Philibert Church in Dijon, France. In *Salt Weathering on Buildings and Stone Sculptures, 22–24 October 2008, The National Museum Copenhagen, Denmark* [Proceedings from the international conference], ed. J. S. Albertsen, 29–40. Lyngby: Technical University of Denmark, Department of Civil Engineering. http://www.design.upenn.edu/files/14-Bourges__Verges_Belmin_Desalination_SWBSS_2008.pdf.

European Commission Research, and P. Jacobs. 2009. SALTCONTROL: Countering Erosion in Europe's Historic Buildings. http://ec.europa.eu/research/fp6/ssp/saltcontrol_en.htm.

Groot, C., R. van Hees, and T. Wijffels. 2009. Selection of plasters and renders for salt laden masonry substrates. *Construction and Building Materials* 23 (5): 1743–50.

Malaga, K., B. Schouenborg, and B. Grelk. 2008. Combadura y dilatación de paneles de piedra natural: Ensayo y evaluación de mármol y caliza [Bowing and expansion of natural stone panels: Marble and limestone testing and assessment]. *Materiales de construccion* 58 (289–90): 97–112.

Price, C., ed. 2000. *An Expert Chemical Model for Determining the Environmental Conditions Needed to Prevent Salt Damage in Porous Materials: Protection and Conservation of the European Cultural Heritage*. Research Report (European Commission, Directorate-General XII, Science, Research, and Development) 11. London: Archetype Publications.

TU Delft (Delft University of Technology). 2009. EU Project Desalination: Assessment of desalination mortars and poultices for historic masonry. http://www.citg.tudelft.nl/live/pagina.jsp?id=267cbaf8-92c8-4204-92c0-97c32fff7eb5&lang=en.

Van Hees, R. P. J. 2002. Project ASSET: Assessment of suitable products for the conservative treatments of sea-salt clay. http://www.onderzoekinformatie.nl/nl/oi/nod/onderzoek/OND1282688/.

Van Hees, R. P. J., S. Naldini, and J. Delgado Rodrigues. 2009. Plasters and renders for salt laden substrates. *Construction and Building Materials* 23 (5): 1714–18.

Selected Stone Conservation Bibliographies

Baer, N. S. 1997. *Materials of Art and Archaeology: Bibliography, Part IX; Stone*. New York: Conservation Center, Institute of Fine Arts, New York University. http://www.nyu.edu/gsas/dept/fineart/faculty/baer/Bibliography9.doc.

Doehne, E. 2003. Building material decay and salt weathering: A selected bibliography. Supplement to *Natural Stone, Weathering Phenomena, Conservation Strategies and Case Studies,* ed. S. Siegesmund, T. Weiss, and A. Vollbrecht. Geological Society Special Publication 205. London: Geological Society of London. http://www.geolsoc.org.uk/webdav/site/GSL/shared/Sup_pubs/2003/SUP18182.rtf.

ICOMOS Centre de documentation = ICOMOS Documentation Centre. 2009. *Stone: Bibliography = Pierre: Bibliographie*. Paris: ICOMOS. http://www.international.icomos.org/centre_documentation/bib/stone.pdf.

ICOMOS-ISCS (ICOMOS International Scientific Committee for Stone) and V. Vergès-Belmin. 2002. Bibliography (Selection). http://lrmh-ext.fr/icomos/consult/index.htm [Bibliography].

Lewin, S. Z. 1966. The preservation of natural stone, 1839–1965: An annotated bibliography. *Art and Archaeology Technical Abstracts* 6 (Suppl. 1): [183]–277.

References from the French Literature in Stone Conservation

Bromblet, P., and G. Martinet. 2002. Joints, mortiers de pose et produits de ragréage: Les différentes pathologies; Réflexions et préconisations. *Pierre actual: Matériaux, ouvrages, techniques* (785): 66–79.

Bromblet, P., J.-D. Mertz, V. Vergès-Belmin, and L. Leroux. 2002. Consolidation et hydrofugation de la pierre. *Monumental: Revue scientifique et technique de la Sous-direction des monuments historiques* (2002): 200–243.

Bromblet, P, and T. Vieweger. 2005. Le laser de nettoyage de la pierre et la restauration des sculptures. *Pierre actual: Matériaux, ouvrages, techniques* (829): 86–95. http://www.lrmh.fr/lrmh/telechargement/laserpb.pdf.

Orial, G. 2005. Les altérations biologiques et les biens patrimoniaux: Introduction. *Monumental: Revue scientifique et technique de la Sous-direction des monuments historiques* 2005 (1): 95.

Orial, G. 2005. Les altérations biologiques et les biens patrimoniaux: Les bactéries, algues et lichens: Morphologie et altérations. *Monumental: Revue scientifique et technique de la Sous-direction des monuments historiques* 2005 (1): 96–99, 117.

Orial, G., and F. Bousta. 2005. Les altérations biologiques et les biens patrimoniaux: Les traitements; Définitions, sélection des produits et mise en oevre. *Monumental: Revue scientifique et technique de la Sous-direction des monuments historiques* 2005 (1): 107–12.

Vergès-Belmin, V., and P. Bromblet. 2000. Le nettoyage de la pierre. *Monumental: Revue scientifique et technique de la Sous-direction des monuments historiques* (2000): 220–73.

Vergès-Belmin, V, and P. Bromblet. 2001. La pierre et les sels. *Monumental: Revue scientifique et technique de la Sous-direction des monuments historiques* (2001): 224–62.

References from the German Literature in Stone Conservation

Siegesmund, S., M. Auras, J. Ruedrich, and R. Snethlage, eds. 2005. Geowissenschaften und Denkmalpflege: Bauwerkskartierung, Natursteinverwitterung, Konservierungsstrategien. *Zeitschrift der Deutschen Geologischen Gesellschaft* 156 (1): 1–238. http://www.schweizerbart.de/papers/zdgg/list/156#paper55362.

Siegesmund, S., and A. Ehling, eds. 2007. Rohstoff Naturstein = Natural Building Stone Resources. *Zeitschrift der Deutschen Gesellschaft für Geowissenschaften* 158 (3): 349–665.

Siegesmund, S., and A. Ehling, eds. 2008. Rohstoff Naturstein: Teil 2 = Natural Building Stone Resources: Part 2. *Zeitschrift der Deutschen Gesellschaft für Geowissenschaften* 158 (4): 667–1087.

Siegesmund, S., and R. Snethlage, eds. 2008. Denkmalgesteine Festschrift Wolf-Dieter Grimm. *Schriftenreihe der Deutschen Gesellschaft für Geowissenschaften* 59: 1–326.

Snethlage, R., ed. 1995. *Natursteinkonservierung I*. Denkmalpflege und Naturwissenschaft. Berlin: Ernst & Sohn.

Snethlage, R., ed. 1998. *Natursteinkonservierung II*. Denkmalpflege und Naturwissenschaft. Stuttgart: Fraunhofer.

Useful Sources of Information for Stone Conservation Research

Free Online Research Databases (arranged in order of usefulness)

Google Scholar (useful for articles):
http://scholar.google.com/

Google Books (some books have searchable full text):
http://books.google.com/

WorldCat (search local libraries and post reviews):
http://www.worldcat.org/

Getty Conservation Institute Art and Archaeology Technical Abstracts (AATA):
http://aata.getty.edu/nps/

Canadian Conservation Information Network BCIN:
http://www.bcin.ca/

ICCROM Library:
http://library.iccrom.org/

The Laboratoire de Recherche des Monuments Historiques (LRMH) has a photographic and bibliographic database called CASTOR:
http://www.lrmh.fr/cgi-bin/qtp?typge=CZIE&lang=uk

The CRNS database CAT.INIST is useful, especially for finding missing abstracts for older articles:
http://cat.inist.fr/?aModele=presentation

The Scirus database by Elsevier is similar to Google Scholar:
http://www.scirus.com/

Commercial Research Databases (by institutional subscription, in most cases; arranged in order of usefulness)

Scopus by Elsevier:
http://info.scopus.com/

Science Direct by Elsevier:
http://www.sciencedirect.com/

ISI Web of Knowledge:
http://isiwebofknowledge.com/

Springer:
http://www.springerlink.com/

Geological Society of London Database:
http://www.lyellcollection.org/

Geological Society of America Publications:
http://www.gsapubs.org/

American Chemical Society:
www.acs.org/

JSTOR Non-Profit Archive:
http://www.jstor.org/

PowerPoint Slides Archive and Network

http://www.slideshare.net/

http://www.slideshare.net/icomos.uk

Online Network of Repositories

http://en.scientificcommons.org/

http://repository.upenn.edu/

Online Building Conservation Education Site
(Practitioner support for building conservation accreditation)

http://www.understandingconservation.org/

Lists of Conservation Related Sites

Getty Conservation Institute List of Sites:
http://www.getty.edu/conservation/research_resources/othersites.html

ICCROM Database of Conservation Related Links:
http://www.iccrom.org/db_links.asp

Conservation Online (Cool) (site formerly hosted at Stanford University, now on a new server at AIC):
http://cool.conservation-us.org/

American Institute for Conservation of Historic and Artistic Works (AIC):
http://www.conservation-us.org/

Robert Gordon University, Aberdeen, UK (links to heritage conservation and related sites [last updated in 2005, but still quite useful]):
http://www2.rgu.ac.uk/schools/mcrg/sites.htm

UK National Conservation Centre:
http://www.liverpoolmuseums.org.uk/conservation/

Forum Restauro @ Conservazione:
http://www.forum-restauro.org/

A variety of documents are available digitally and can be found via their DOI (digital object identifier). Similar to URLs, DOIs do not change as Web sites change. When the DOI for a document is known, the document can be located by accessing a DOI resolver, such as http://dx.doi.org, and entering the DOI.

Index

About the Authors

Eric Doehne holds a BS in geology from Haverford College (Pennsylvania) and MS and PhD degrees in geology from the University of California, Davis. He joined the Getty Conservation Institute in 1988 and worked there as a research scientist until 2010. Doehne's projects included a wide range of research collaborations, including the desalination of historic masonry in New Orleans, the conservation of magnesian limestone buildings in Yorkshire, and coral-red gloss on Greek vases. In 2010 he founded Conservation Sciences, LLC—a consultancy focused on materials science for art, architecture, and archaeology. He specializes in consulting, teaching, and investigations about the composition, behavior, and treatment of such historic inorganic materials as stone, glass, and ceramics. Doehne's particular interest is in stone decay mechanisms, for instance, the role of soluble salts in the deterioration of architecture, wall paintings, and sculpture.

Clifford Price is Emeritus Professor of Archaeological Conservation at University College London (UCL). He earned an MA and PhD in chemistry at St. Catharine's College, Cambridge, and worked first at the Building Research Establishment. In 1983 he was appointed head of the Ancient Monuments Laboratory with English Heritage, and in 1990 he moved to the Institute of Archaeology, UCL. His research has been concerned primarily with the decay and conservation of stone. He has been involved in the conservation of several English cathedrals and has undertaken consultancies at sites around the world. He retired from UCL in 2007.